The
Science
and the
Myth
of
Melanin
Exposing The Truths

T. Owens Moore, Ph. D.

EWORLD INC.

Buffalo, New York
14209
eeworldinc@yahoo.com

To my second son, Jabari Adisa
he was born out of the womb of darkness
on October, 15th, 1994,
in the same year
that I completed this book.

ISBN 978-1-61759-016-0

Formally published by
A&B Publishers Group
Brooklyn, New York
ISBN 1-886433-97-6

Published by

EWORLD INC.

Buffalo, New York
14209
eeworldinc@yahoo.com

COVER CONCEPT: EWorld Inc.

Manufactured & Printed in the United States of America

10 11 12 13 14 15   6 5 4 3 2

Contents

Part Two--Melanin Physiology

Acknowledgments

To my brother, Jelani Makalani Madaraka, for his consistent help in editing and formulating ideas. Your incalculable hours of assistance will always be cherished.

To many African-centered soul searchers such as Dr. Niyana K.B. Rasayon for stimulating interest in this area of research.

To my wife Terrie and my son Tyehimba for their patience in allowing me to complete this document during precious family time.

I will be forever grateful to my supporting and trusting family members, George and Margaretta Bonds and Walter and Mara Moore. Your assistance has made this project a reality. Thank you for your confidence.

Lastly, I must acknowledge the professional assistance extended by Beckham House Publishers, Inc. Barry Beckham, thank you for believing in this project. It was an honor to work with you, your staff and all artists and editors responsible for completing this work. African-American writers are fortunate to have you as a quality African-American-owned publishing company. I wish you the best in this competitive field.

Preface

S cience is the systematic exploration of natural phenomena. It is the best mode for investigating phenomena that are either complex or not well understood. With scientific facts, one is better able to explain the characteristics of the physical world. Melanin, the controversial subject of this book, is a highly complex molecule. Consequently, information on melanin is often expressed in confusing scientific jargon. Therefore, people may extract some information about melanin functioning yet not understand "the science of melanin." I have written this book in order to demystify and clarify melanin research and to demonstrate scientifically why melanin is necessary for optimal human health.

Melanin is a substance found in practically all living organisms. It is also found in inanimate objects as small as a piece of fruit and as large as celestial bodies. In this book, however, we are concerned specifically with the functioning of melanin in the human body. We will review scientific research on melanin to explain its role in human development, physiology, and behavior. The content relies on scientific evidence to ensure that the book not be classified as "pseudoscience." Although I have not written the book as a defense of melanin research, *The Science of Melanin* is indeed a response to melanin critics.

There is a common misconception that all melanin researchers are "racists" or believe skin color, or degree of melination, is responsible for psychological and behavioral differences between black and white people. While such notions obviously are false, it is easy to see how some melanin scientists could unfairly be labeled "racist." Historically, scientists espousing racist viewpoints were considered "biological reductionists," persons who reduce all behavior to biological phenomena. If individuals present information pertaining to the "superiority" of melanin functioning but offer no substantive evidence, one can see how melanin researchers might be tagged also as biological reductionists. Rather than fall into these traps of obfuscation, we will explore the realm of melanin functioning, in scientific as well as human terms, relying only on scientific fact. Only in this way will we controvert the critics of and learn more about the exciting field of melanin research.

Introduction

I am a scientist dedicated to providing positive information about people of African descent. To help you better appreciate my dedication to this important task, I will describe my academic background.

As an undergraduate at Lincoln University in Pennsylvania from 1982 through 1986, I began my quest to become a scientist. I entered as an engineering student. After a very stimulating psychobiology course, however, I decided to major in psychology. Nagged by a lingering interest in pure science, I enrolled as a pre-medical student. Then, with a B.S. in psychology, I planned on attending medical school to focus on psychiatry or neurology. But after leaving Lincoln, I was not accepted to medical school. My advisor at Lincoln, Charles Duncan, suggested graduate school at Howard University in Washington, D.C. I attended Howard from 1987 to 1992.

Thanks to Dr. Duncan's guidance, I have completed my graduate career, earning an M.S. degree in 1989 and a Ph.D. in 1992, both in physiological psychology. Sadly, Dr. Duncan made his transition to the spiritual world on June 9th, 1994. Since Dr. Duncan is not here physically to read my words of gratitude, I would like his family to know that Dr. Duncan has touched the lives of many students. Personally, I can say he has had a positive influence on my academic career.

My nourishing educational experiences at Lincoln and Howard provided the opportunity to study under many strong African-American scholars. The list is long. However, I believe it is appropriate to mention them in gratitude: Drs. Duncan, Leslie Hicks, Jules Harrell, Alphonso Campbell, Albert Roberts, C. Ovid Trouth, Hope Hill and Mrs. Anita Moore-Hackney. In their own ways they instilled in me a concern for excellence, and through this book I have attempted to sustain their legacy.

In the footsteps of profound men such as Kwame Nkrumah and Thurgood Marshall, who both attended Lincoln and then Howard, I am continuing the journey. My rich experiences at Lincoln and at Howard provided me with the mental strength and stamina to cope in a scientific world controlled by white scholars and administrators. While attending Howard, I worked at Walter Reed Army Institute of Research from 1989 to 1992. In this research environment I gained tremendous experience from renowned scientists like James Meyerhoff and Michael Potegal. Meyerhoff is an expert in psychoneuroendocrinology, and Potegal is an expert in the study of aggressive behavior. Among my many scientific interests, I conduct research pertaining to the experiences I gained at Walter Reed. I am an assistant professor in the department of psychology at Morehouse College in Atlanta, Georgia where I teach and conduct research.

I also had the fortunate opportunity at Howard to spend hours reading literature and attending lectures pertaining to African history and culture. In the course of my search for knowledge of self, the topic of melanin kept surfacing. Since my area of specialty was brain and behavior relationships, I attempted to make sense out of the literature and presentations on melanin. I came to the realization that much of the information on melanin from an African-centered perspective was misleading. With the intention of elucidating this most important and complex topic, I have studied melanin since 1987. Seven years later my ideas have culminated in this book.

I encountered numerous shallow and superficial critiques of melanin research. The misinformed and often ignorant criticism came from white and nonwhite critics alike. Interestingly, the harsher attacks have been launched mostly by people with little scientific preparation. I felt it to be my duty to provide serious information on the subject. Many of my elders and scholarly mentors suggested that I stay away from the topic, lest I be labeled a "pseudoscientist." However, I felt a strong call to respond to the intellectual world. So, as a conscious African living in America, I have made it my work to present this document on melanin. If African-centered melanin researchers make no legitimate responses to the vicious attacks upon this most important field of study (Ortiz de Montellano, 1993), African-centered scientists will receive yet more inappropriate negative publicity.

Unfortunately, the controversy over melanin research has intensified. The topic has received considerable attention from the white media. For example, author Leon Jaroff (1994) wrote an excoriating article, "Teaching Reverse Racism" in *Time* magazine. The underlying message is that most black scholars who teach black studies are using a strange doctrine of "black superiority" when they discuss melanin. Essentially, Jaroff believes... "[i]n a society that has treated blacks as inferior because of the color of their skin, it is hardly surprising that many of them now embrace melanist doctrine. But in doing so, they are indulging in what they have long decried: racism." Those who run the white power structure commonly use reverse rhetoric to castigate the self-affirming viewpoints of African-centered scholars. In other words, whites project their own racism onto us in order to avoid their insecurities. Henceforth, many insecure white people believe that any Black-African consciousness, history, or related information is "racism" in reverse. I have not written this book to persuade someone who may be conceptually, shall we say, incarcerated. African and African-American scholars must go beyond trying to change the mentality of people who have been indoctrinated from birth to believe that people of European descent are superior. The open-minded person who is able to read objectively will learn the most from this book. The close-minded, who maintain *a priori* conclusions that melanin is a farce, will miss out on a learning experience. Since some people know nothing about melanin nor care to, I hope this book will serve as a resource to stimulate interest and "endarken" the world on the physiological importance of melanin functioning.

Here at the outset, it should be emphasized that the human body is not just two legs, two arms, a trunk and a head. The human body is a vehicle that can carry human beings toward optimal physical and spiritual health. Western science does not study or teach that the human body can be a vehicle for this transformation. The primary focus of Western science is control. Generation after generation of physicians are trained to master their patient's life through external control; control mechanisms like drugs, psycho-therapy, hospitalization, and surgery are the mainstay of Western medicine. Through this process of control the medical-pharmaceutical-industrial complex grows rich while patients become poor. To improve individual health, more emphasis should be placed on prevention. A responsible and efficient medical system advises individuals on how to avert illness, not on how to besiege a patient only after an illness develops. By enriching your body through diet, nutrition, meditation and positive thinking, you can avoid illness and maximize your health. We will look at how melanin can help.

In the black community, many scholars make reference to melanin in their speeches. Most of these scholars echo a similar message that melanin is "the chemical key to black greatness" or "the hidden factor in the health, well being and creativity of black people." Such statements are highly polemical, mainly for two reasons. First, the majority of speakers comment on melanin functioning though they have performed no meaningful research themselves. The most convincing and acceptable way to discuss such topics is to provide documented research, thereby providing readers with valid and useful information.

Secondly, there are many factors contributing to behavior; melanin alone can hardly be considered the source of the good as well as the bad behavior in the Black community. The current level of crime, poverty, and decadence among Black people more accurately indicates that there are social, cultural, and moral factors that can overwhelmingly influence a people. Certainly, melanin scholars are aware of the necessity of scientific data as well as the cultural realities that help form people's lives. Nonetheless, audiences are often misguided amid the volatile melanin controversy, and this book is for them.

In Chapter one, we will critically analyze five books: 1. *The Isis Papers* by Frances Cress Welsing; 2. *Melanin: The Chemical Key to Black Greatness* by Carol Barnes; 3. *Jazzy Melanin: A Novel* by Carol Barnes; 4. *Color Me Right...Then Frame Me in Motion* by Malachi Andrews; and 5. *The African Origin of Biological Psychiatry* by Richard King.

To my knowledge, the documents herein critiqued are the most popular available to the public. They explain melanin from an African-centered perspective. As I will demonstrate, these texts are deceptive because they lack a solid foundation concerning the science of melanin.

Collectively, the general theme found in the viewpoints of scholars such as Welsing, Barnes, Andrews and King has led many readers to believe that melanin alone is the entity that makes black people psychologically and behaviorally different from non-black people. Given the complex nature of human behavior, a person is defined by much more than biology. Anyone who has critically analyzed and scientifically researched melanin formation will find the viewpoints held by these scholars to be insufficiently supported.

Before we proceed, let me state that any study of melanin is technical by nature, and there is no way in which this topic could be thoroughly analyzed without using complex scientific terms. Therefore, the following paragraphs reveal the technical nature of this area of research. No apology is made for subjecting you to this terminology in the introduction because it is more important that we deal with the scientific terms to prove that this document is not "pseudoscience." How is melanin formed? There are many types of melanin associated with the human body and the surrounding environment. It is inappropriate to categorize skin melanin, visceral melanin, melanin in digestible food items, and melanin in the sky as the same. From this complex and diverse discussion on melanin, the implied message or most common association is that skin melanin is the particular type of melanin that influences behavior. It is erroneous to assume that the physical attribute of skin pigmentation (melanin) affects a person's mental or brain capacity. Skin pigmentation has nothing to do with one's intellectual abilities. Although there are similar bioelectronic properties associated with skin and brain melanin, melanin is synthesized and formed differently in the skin and brain. Moreover, there are different enzymes and biochemical processes that determine the formation of the many melanins that exist in and outside of the body. It is mere speculation, however, on how melanin is formed in the cosmos.

The essential point we wish to convey is that skin color is not directly related to their behavioral capacity to function in society. Since brain melanin is attributed to behavioral functioning, it should be emphasized that brain melanin is the type that can be said to influence behavior. It is my opinion that brain melanin is genetically programmed to function at different capacities depending upon a person's overall capacity to produce melanin.

In other words, every gene manufactures a specific chemical substance for the proper functioning of the organism. An excess or deficiency of a certain chemical (e.g., enzyme, amino acid sequence, protein) could adversely influence the organism's ability to function normally. The specific chemicals needed to biosynthesize melanin are genetically programmed to function at a different capacity in people with a darker hue. Empirical evidence for the difference can be observed in two ways: 1) the amount of external pigmentation; and 2) the enhanced psychomotor capabilities that stem from the functioning of melinated brain structures in the motor system.

Another point about melanin that requires more clarity is the belief that dopamine is a precursor molecule leading directly to brain melanin formation. (Dopamine is a neurotransmitter classified as a catecholamine.) Although there is evidence supporting as well as refuting the hypothesis that dopamine is involved in the biosynthesis of melanin, the issue remains open. A major error made by people who think that high amounts of brain dopamine produce high amounts of brain melanin is the association of dopamine and melanin with the mental disorder known as schizophrenia. Schizophrenia is a type of psychosis characterized by disordered cognitive functioning and poor social adjustment due to specific brain malfunctions. Some of the symptoms include hallucinations, delusions, distorted reality, and abnormal changes in brain structure. It so happens that one biochemical maladjustment in schizophrenic patients is increased levels of dopamine.

Since dopamine and melanin coexist in high concentrations in a brain site called the substantia nigra, it is commonly believed that dopamine is directly involved with melanin formation in the brain. Some African-centered melanin researchers imply that high amounts of dopamine in the brain lead to higher amounts of melanin in the brain. To believe that high amounts of dopamine are required to elevate brain melanin in black people would suggest that highly melinated black people should be classified as having schizophrenia. Such a notion is clearly nonsensical.

Another viewpoint that requires additional clarity is melanin's proposed biosynthesis from neurotransmitters such as dopamine, norepinephrine and epinephrine. Each of these neurotransmitters is dependent upon a specific enzyme to convert it into another chemical structure. In general, these neurotransmitters are found in separate locations in the brain and influence specific brain functions. In other words, dopamine, norepinephrine and epinephrine all have different functional roles as neurotransmitters. Enzymes are the substances responsible for converting each of these neurotransmitters into other chemical structures.

Barnes (1988) makes two points in his book to suggest that neuromelanin is probably formed by dopamine. First, he hypothesizes that "dopamine may be converted to dopamine-quinone probably utilizing the tyrosinase catalyst." Tyrosinase would be acting as an enzyme in both the brain and skin; however, this dopamine-quinone formation is normally associated with skin melanin, and the process can be found in the Mason-Raper Pathway (see Figure 1). Secondly, he suggests that "dopamine may proceed through steps involving norepinephrine and epinephrine and eventually polymerizing into neuromelanin." I question this latter viewpoint because physiologically there is no clear evidence linking catecholamine synthesis and neuromelanin formation.

Barnes is only speculating on how neuromelanin is formed. If dopamine is converted into neuromelanin by a tyrosinase catalyst, then there should be high enzyme activity in the substantia nigra and other pigmented brain regions. Rodgers and Curzon (1975) used an elaborate experimental technique, radiometric assay, to provide substantial evidence that there is no enzyme activity in the substantia nigra or any other pigmented brain regions that were investigated in human brain tissue. The authors concluded that brain melanin formation may be a largely non-enzymic process. Therefore, focusing on some hypothetical enzymic process for brain melanin formation and surmising that high amounts of a neurotransmitter such as dopamine are specifically associated with the brains of black people can be misleading. It may be more prudent to state simply that brain melanin is genetically programmed to function at a different capacity depending upon a person's overall genetic capacity to produce melanin.

In reference to his second point, Barnes makes a suggestion that dopamine and other neurotransmitters may polymerize into brain melanin. This polymerization is highly dependent upon enzyme activity in highly melinated brain sites such as the substantia nigra. Since enzyme activity in the substantia nigra has not been substantially validated through experimental research, less emphasis should be placed on some enzymatic process that eventually polymerizes neurotransmitters into brain melanin. Although it is an unresolved issue, it may be more appropriate to state, yet again, that brain melanin in the human species is dependent upon genes. Skin melanin, also, is dependent upon genes, but enzyme activity is highly associated with skin pigmentation and not with the formation of brain melanin.

Furthermore, we can observe that levels of skin melanin are not positively correlated with the presence of melanin in the brain. The simple fact that albinos, who lack skin melanin, have no abnormal change in substantia nigra pigmentation indicates that there is no direct correlation between the amount of melanin in the skin and the amount of melanin in the brain. A special character of brain melanin is suggested by the normal pigmentation of the substantia nigra and locus coeruleus of albinos, who lack melanin pigments elsewhere (Foley and Baxter, 1958), and by the absence of reports of melanomas of these brain regions even though melanomas of other melanized tissues are well known (Curzon, 1975).

I do support the perspective, however, that the capacity for brain melanin functioning is greater in people of African descent, a view supported by genetics research. For example, substantia nigra pigmentation was not found in rabbits, guinea pigs, or most other marsupials, but there was noticeable pigmentation in cats and dogs. In addition, the substantia nigra has been shown to be slightly pigmented in laboratory rodents. These findings indicate that there are species differences in neuromelanin-containing cells. These differences are also found in primates and man, and there are differences according to age distribution in man (Mann and Yates, 1983; Marsden, 1983). Accordingly, one can logically surmise that there are ethnic differences as well, i.e. black versus white, in neuro-melanin-containing cells.

In sum, it is recommended that researchers make a distinction between skin and brain melanin. Moreover, one should validate scientifically any hypotheses suggesting that melanin can enhance one's mental capabilities. According to the information available to this scientist, melanin is genetically programmed to function according to an individual's overall genetic makeup.

In this book, we want to emphasize that our physical bodies contain all of the equipment necessary to enhance our physical and mental well-being. We will review research that explains melanin's ability to organize our cells from the earliest stages of embryological development, maximize the performance of our physiological systems, and prolong our lives. We will review scientific articles on melanin research. This book's main goals are not only to capture the attention of scientists who want to pursue melanin research, but also to inform nonacademicians.

It is divided into three parts. Part One begins with a critique of literature written by African-centered scholars. We want to provide the reader with background information on how other scholars view melanin. Also, a general explanation of the location and distribution of melanin will be mentioned with emphasis on the origin of melanin from the cellular level. This general explanation provides the groundwork for Chapter three, where we probe more deeply into the advantages of melanin functioning. Part two consists of specific roles that involve melanin. Separate chapters detail the specific roles for melanin functioning in the nervous, endocrine, visual, auditory, and vestibular systems, and in early childhood development. Part three speculates on the metaphysical implications of the scientific phenomena of melanin. It is proposed that melanin is a link between the material and spiritual realms of existence. We conclude with ideas for prospective research.

T. Owens Moore
Decatur, Georgia
February 16, 1994

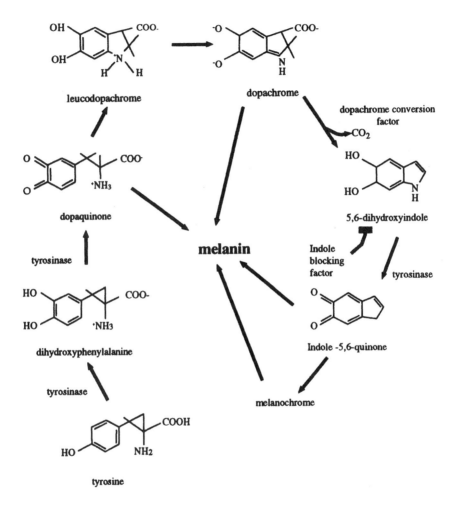

Figure 1. The Mason-Raper Pathway

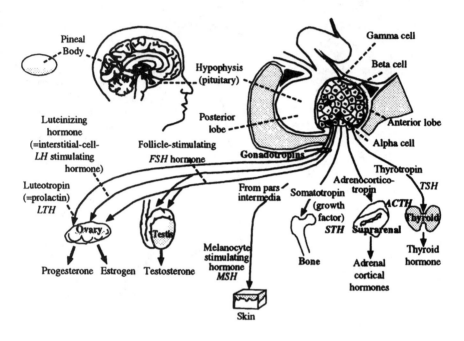

Figure 2. The adenohypophysis produces several hormones controlling the activity of a number of endocrine glands.

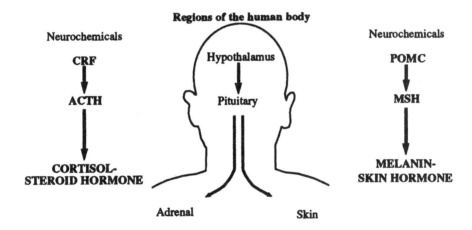

Figure 3. Melanin stimulation functions similarly to other endocrine hormones such as the steroid hormone cortisol. In a cascade of physiological events starting in the brain and leading to the body site, numerous neurochemicals regulate the functioning of hormones. For example, the specific regions of the hypothalamus secrete both CRF and POMC. These neurochemicals stimulate hormones in the pituitary such as ACTH and MSH. ACTH stimulates the adrenals to produce cortisol, and MSH stimulates the release of melanin in the skin.

Melanotropin Source	Site	Physiological response
Pars intermedia-a-MSH	Propigment (stem) cells	Melanoblast cytodifferentiation
	Epidermal melanocytes	Melanin biosynthesis (morphological color change): skin, hair, and pelage pigmentation
	Dermal chromatophores	Physiological color change
	melanophores	melanosome dispersion
	iridophores	reflecting platelet perinuclear aggregation
	xanthophores/erythrophores	carotenoid and pterinosome dispersion
	Sebaceous (dermal) glands	Sebum production
	Preputial glands	Lipogenesis and pheromone production
	Fetal adrenal cortex	Steroidogenesis (fetal growth and development?)
	Unknown	Stress adaptation
Pituitary melanotropin	Adrenal zona glomerulosa	Aldosterone synthesis and secretion
Brain neurons (α-MSH or [desacetyl]-α-MSH	Other brain neurons	Inhibition of opiate peptide induced analgesia
		Neurotransmitter or neuromodulator functions: arousal, attention, learning, memory retention
		Central temperature control
	Anterior pituitary	Stimulation of ACTH secretion
		Stimulation of STH secretion
Pineal gland (epiphysis cerebri)	Unknown	Circadian processes of the pineal (rat)

Figure 4. Extrapigmentary effects of melanocyte stimulating hormone

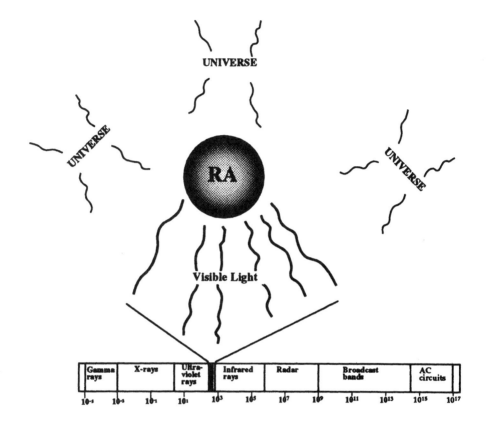

Figure 5. Electromagnetic Spectrum

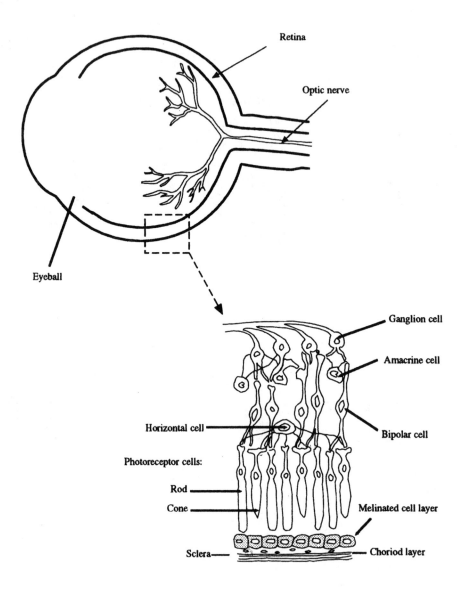

Figure 6. Diagram of the eye and pigment layer of the retina.

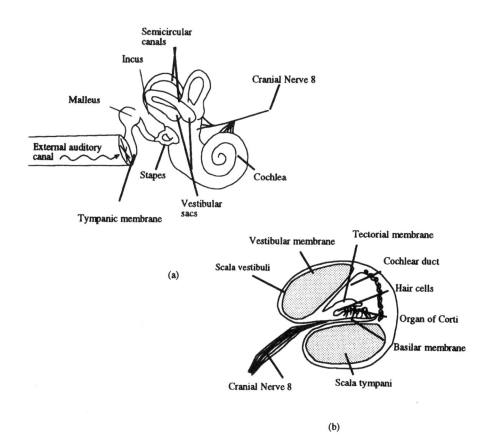

(a)

(b)

Figure 7. Diagram of the ear

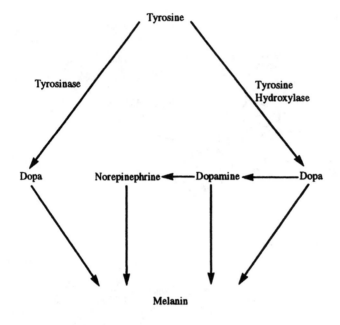

Figure 8. Alternative ways to form brain melanin from biogenic amines

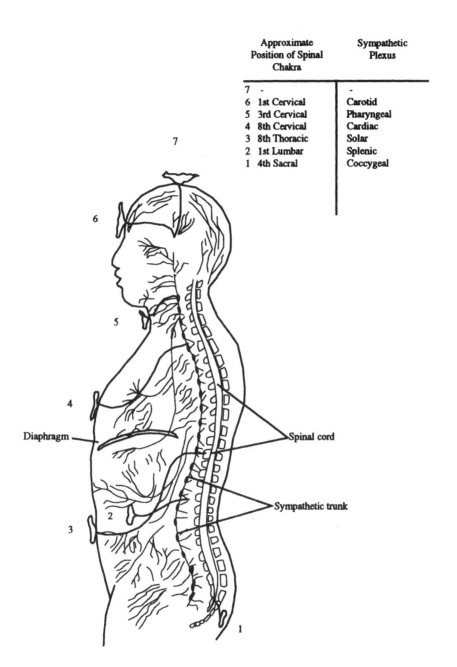

	Approximate Position of Spinal Chakra	Sympathetic Plexus
7	-	-
6	1st Cervical	Carotid
5	3rd Cervical	Pharyngeal
4	8th Cervical	Cardiac
3	8th Thoracic	Solar
2	1st Lumbar	Splenic
1	4th Sacral	Coccygeal

Figure 9. Representation of the chakra system in the human body

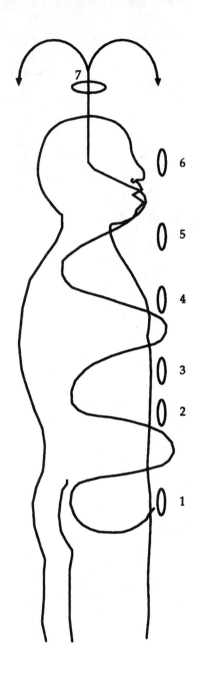

Figure 10. Energy flow up the chakra system

Figure 11. Seven planes of existence, the human body and the link with Divine Universal Consciousness

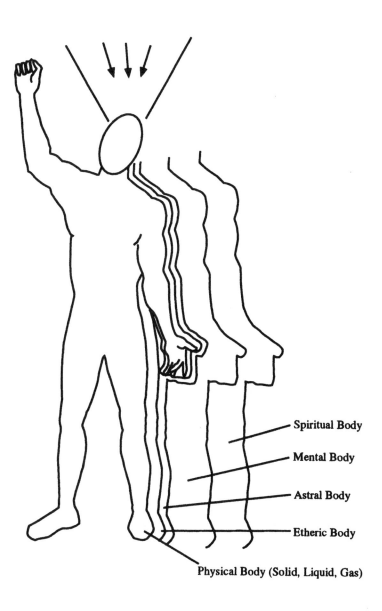

Spiritual Body

Mental Body

Astral Body

Etheric Body

Physical Body (Solid, Liquid, Gas)

	Name of Chakra	Sanskrit Name	Endocrine Gland
7	Crown	Sahasrara	Pineal
6	Brow	Ajna	Pituitary
5	Throat	Vishuddha	Thyroid
4	Heart	Anahata	Thymus
3	Navel	Manipura	Pancreas
2	Spleen	Svadhisthana	Adrenals
1	Root	Muladhara	Gonads

Figure 12. A table that correlates chakras with body organs.

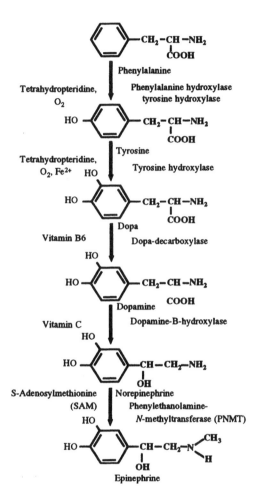

Figure 13. The biosynthesis of neurotransmitters and the role of vitamins

Part One - Overview

1

Critical Analysis:
Four African-American Writers

Many different perspectives have been used to criticize African-centered melanin researchers. And many of the harsh critics have been black intellectuals far removed from the hard sciences. Unquestionably, there are some serious flaws in a few melanin presentations that have been given by black scholars. In this chapter, we will address some of these flaws in an attempt to provide some clarity to melanin research. We will not "argue" against melanin theorists. We will, however, critically analyze some of the prevailing perspectives in order to clarify important concepts. Moreover, this critical analysis will include an up-to-date review of the most recent African-centered literature on melanin.

OVERVIEW

Cheikh Anta Diop (1923-1986), born in Diourbel, Senegal, was one of the greatest and boldest African scholars to live in the twentieth century (Van Sertima, 1987). He was a multi-talented researcher in the disciplines of anthropology, archaeology, ethnology, linguistics, history, chemistry and physics. Diop used these disciplines to correct the Eurocentric view of world history. The task, however, was not simple. He broke through tremendous political barriers to obtain a Ph.D. from the University of Paris. His independent research opposed all that had been taught in Europe for two centuries about the origin of civilization, so his white mentors did not support him. In a demonstration of his perseverance, he wrote more than one dissertation to earn, finally, the doctoral degree.

The themes emanating from Diop's early work are threefold:

1. Egypt was the center of Africa's main cultures and languages.
2. Egyptian civilization received its main influence from the central part of Africa.
3. The creators of classical Egyptian civilization were black-skinned people of African descent.

As a scientist, he used his training to validate this last theme that Egyptian civilization was African-based to support his other research. As the founder and director of the radiocarbon laboratory at the University of Dakar, Diop (1973) developed a chemical process for testing the melanin content in the skin of Egyptian mummies to establish their black African ancestry. Unfortunately, for political reasons, he was unable to conduct fully his research. Had he been free to conduct his work, he could have provided the scientific world with crucial research on melanin.

It is unclear whether Diop was even interested in the science of melanin *per se* since he presented no argument that melanin had any influence on behavior. His primary intention in studying melanin was to prove that the ancient Egyptians were black-skinned. Any differences in black and white behavior, according to Diop, were the result of climatic and other environmental variations. His position was firm on this latter point:

> "...the Black man must become able to restore the continuity of his national historic past, to draw from it the moral advantage needed to reconquer his place in the modern world, without falling into the excesses of *Nazism in reverse* for, insofar as one can speak of a race, the civilization that is his might have been created by any other human race placed in so favorable and so unique a setting" (1976, p. 235, italics mine).

In other words, if white people had been in the Nile Valley while black people were stuck in the ice age, today's world history might be reversed. Diop does not, therefore, express the view that we should label biology as a factor that contributes to the development of civilizations. Rather, he postulated that we should not repeat the fallacious arguments about biology and behavior espoused by Hitler's regime and carried out in brutal experiments by Nazi "doctors."

In sum, Diop strongly believed that "Africa's soul had been stolen and could only be retrieved through a scientific approach" (Van Sertima, 1987). His scholarly work is a model for all scientists to follow. If alive today, he would probably not support a biological reductionist view of melanin as the motivational force behind people's actions. According to his own statements, Diop would support the notion that the environment is the causative agent. Frances Cress Welsing, however, supports the viewpoint that melanin plays a key role in people's behavior.

From a different perspective and continent, Frances Cress Welsing developed some intriguing ideas on melanin. Although we will discuss Welsing's perspective in the next section, she is mentioned in this overview because her views have had a significant impact on other black melanin theorists.

In 1969, she conceptualized a provocative theory called the Cress Theory of Color-Confrontation. Essentially, Welsing believes that white people have developed and currently maintain a global white supremacy system because of two main points that relate to melanin: 1. white people feel genetically inferior because they lack melanin;

and 2. white people realize they are a minority population and numerically inadequate. According to Welsing, these two factors drive the genocidal behavior of white people toward nonwhite people. Her theory has received considerable attention over the years, and she has provided the earliest African-centered literature on melanin.

Before the formulation of Welsing's theory in 1969, many scholars from various disciplines convened on the West Coast of the U.S. and organized "think tanks" to discuss melanin functioning. In California, the Melanin Think Tank officially organized and meetings alternated between Los Angeles and the San Francisco Bay area (Warnette and Andrews, 1990).

Several major think tanks were convened in the mid 1980's, and organizers came together to hold the First Annual Melanin Conference (1987) in San Francisco. The Melanin Group organizers were Richard King, Hunter Adams III, Neferkare Stewart, Carol Barnes and the late Malachi Andrews. The group now includes more members, known collectively as KM-WR Science Consortium. An annual Melanin Conference is held in different regions of the U.S. to discuss melanin research, and the conference represents an arena for the development of African-centered ideas.

The eighth annual conference (1994) was held in Washington, D.C. Since the inception of the 1987 conference, several documents have been written by African-American scholars. What follows is a critique of works by Welsing, Barnes, Andrews, and King.

FRANCES CRESS WELSING

Welsing, a practicing psychiatrist in Washington, D.C., has been a major contributor to the topic of melanin for many years. Her original theory has been republished in her book, *The Isis Papers: The Keys to the Colors* (1991) which is a collection of her essays written over 23 years during her practice. Many of her ideas are derived from therapy sessions conducted with her clients, and her analysis does not document any scientific research on melanin.

Welsing's ideas on racism or white supremacy develop from the work of Neely Fuller (1969). In addition, after graduating from medical school, she went on a "fact finding mission" to Nazi Germany to see if she could uncover the mentality that allowed genocide against the so-called Jews. These two influences coupled with her training molded her thoughts for the Cress Theory of Color-Confrontation.

Welsing believes that whites have developed and maintained a global system of supremacy to prevent their genetic annihilation. Because whites lack the genetic capacity to produce significant levels of melanin, the global white minority must act genocidally against people of color in order to ensure white genetic survival. This "kill or be killed" mentality, says Welsing, "is the reason that persons who classified themselves as 'white' behaved genocidally towards semites in the holocaust in Nazi Germany Europe (1933-1945)." On another historical note she adds, "This is also the reason that persons who classified themselves as 'white' behaved (and still behave) genocidally towards the indigenous inhabitants of the Western Hemisphere who were classified as red (non-white)."

As a psychiatrist, her thinking has tended to focus on describing the motivations that have created the white supremacy system. She has reasoned that in many cases, neurotic drives for superiority and supremacy are usually founded upon a deep and pervading sense of inadequacy and inferiority. According to Welsing, the lack of melanin is the propelling force for these drives. The core of her psychogenetic theory is stated in the following passage:

> "The Theory of Color-Confrontation states that the white or color-deficient Europeans responded psychologically with a profound sense of numerical inadequacy and color inferiority upon their confrontations with the massive majority of the world's people all of whom possessed varying degrees of color producing capacity. This psychological response, be it described as conscious or unconscious, was one of deeply sensed inadequacy which struck a blow at the most obvious and fundamental part of their being, their external appearance." p. 4.

In effect, this proposed psychological reaction of whites has been directed toward all peoples with the capacity to produce the melanin skin pigments, and "the most profound aggressions have been directed towards the black, 'non-white' peoples who have the greatest color potential and therefore are the most envied and the most feared in genetic color competition."

In any scientific or theoretical analysis, the greater a theory stands against the tide of facts that attempt to refute it, then the greater the theory. Welsing's points about the white supremacy culture are logical and extremely difficult to debate. And it is difficult to ignore the human destruction caused by certain groups of Europeans (Portuguese, English, French, Spanish, etc.) in the last 500 years. For practical implications, Welsing's theory is a rational basis for non-white peoples throughout the world to understand the motivational nuances of individual and collective white behavior. In the words of Welsing:

> "The Color-Confrontation thesis theorizes that the majority of the world's people, non-whites, were manipulated into subordinate positions because never having experienced such a state in terms of their own thought and logic processes and premises, they were unprepared to understand patterns of behavior predicated upon a sense of color deficiency and numerical inadequacy." p. 12.

Welsing should be commended by the world's community for boldly presenting such a provocative theory. Her goal was and is not to spread hate, but to analyze the white supremacy epidemic and neutralize it with rational and logical treatment. Although her theory is a valiant attempt to explain complex human behavior, the analysis has several theoretical pitfalls.

For example, St. Clair Drake (1987) critiqued Welsing and other melanin theorists and found faults in their analyses. There are three main criticisms formulated by Drake that explain his position. First, he states that Welsing's views are an already validated proposition

8

without testing. Since she did not perform any surveys or collect data by conducting experimental research, Drake is saying that her theory should be testable for it to be given credibility. One of the requirements of any theory is that it must be applicable to research. If a theory is not testable, it is essentially useless. True, one should never draw conclusions without proper experimentation. But Welsing acknowledges that her views stem from case studies with patients and from reflections on the historical record of atrocities by some Europeans (e.g., genocide against African people, aboriginal people, and white Jews). Since Welsing was only attempting to analyze white behavior with a conceptual theoretical framework, we can understand why she may not have subjected her theory to the rigor of experimental research.

Secondly, Drake believes Welsing has made an unresearched and therefore unsupported value judgment about biology. For example, Welsing theorizes that a lack of melanin in white people is the biological factor contributing to most of the problems in today's world. This is akin to suggesting that biological factors motivate people's actions. And she offers no data to support this provocative but misleading notion. There is no known evidence linking a lack of melanin with the collective behavior of white people toward nonwhite people. Following Welsing's unsubstantiated reasoning on melanin, one would have to conclude that high melanin levels are the cause of the collective behavior of black people world-wide.

In terms of the third criticism, Drake states that melanin theorists are turning the clock of history backward to biological determinism. It was thought in the last few hundred years that black people were "unintelligent" and "animal-like" because they were black in color (see Gould, 1981). Many Eurocentric and racist scientists were devising theories to prove their belief that blacks were inferior. Drake believes that melanin theorists are falling into the trap of earlier scientists who were setting out to prove *a priori* assumptions.

By reflecting on the past, we see that biological determinism leads one down a blind path. If melanin was the key factor for the genocidal behavior of white people towards non-white people, then what explains white on white crime that has always existed? What explains the fact that white people have massacred their own during various phases of history? To explain this point more clearly, suppose another theoretician invented a random motivational factor--hair texture, for example. In other words, the scenario would be that Europeans have collectively mobilized to fight off the intrusion of kinky hair into their gene pool. Therefore, the motivational factor of hair texture would be pitting one against kinky-haired people. Such a "theory" may seem preposterous, but the "theoretician" may actually find information that appears to defend it.

According to the Cress Theory, one may give too much credit to white people for their organizational abilities. Given the idea white people have been so effective in manipulating human behavior to their advantage, some individuals interpreting the Cress Theory may be led to subconsciously believe that white people have a psychological advantage and should inevitably rule the world. Moreover, one may believe that white people world-wide sat down to devise a plan to maintain global supremacy. The Berlin Conference of 1884 was certainly organized with the goal of controlling the world's resources, but can we really say that these white leaders consciously acknowledged that they lacked melanin, and this predicament motivated them to unite against the people of Africa? No, the motivation behind the Berlin Conference was primarily economic.

Another problem with Welsing's perspective is the association of melanin with either good or evil. Although the historical record demonstrates that some white people have unquestionably been evil and destructive on the world scene, there have been different perpetrations of evil demonstrated by many ethnic groups. We can say from the historical record, however, that evil has reached epidemic proportions in white "civilizations." [We have put civilizations in quotes because it could be called barbarism (Diop, 1991)]. Since highly melinated black people have also been "evil," it is difficult to label melanin the key factor influencing good and evil behavior.

Welsing's position on the neurochemical basis of evil (1991), first formulated in 1988, is stated accordingly:

"...that the absence of this black pigment in the skin

and other aspects of the nervous system - critically impairs the depth sensitivity of the nervous system and the ability to tune into the total spectrum of energy frequencies in the universe."

This African-centered statement about melanin is validated by some research that will be reviewed in Chapter three of this book. However, the next part of Welsing's perspective is not as easy to validate. She continues:

"This deficiency of sensory awareness sets the stage for the absence of harmony (the chaos and destruction), which is evil. Thus, the injustice and evil of white supremacy not only has its foundation in numerical minority status of the global white population and its genetically recessive status in terms of melanin pigment production, but the very absence of melanin in the nervous system in significant degrees (decreasing sensory input and thus sensitivity) is an additional contributing factor in the problem of white supremacist injustice." p. 238.

This passage is misleading. It implies that the absence of melanin in the nervous system in significant degrees is the cause for evil behavior in white people. In other words, she theorizes that the psychological and social behavior of white people is biologically driven. Welsing has made a point to fit her theoretical position. The above passage can lead the reader to assume that all whites are evil and all blacks are good. She does not state this, but the message implied is similar to the Nation of Islam doctrine that all white people are devils. It would be more logical to conclude, whether in scientific or theoretical terms, that every human being has potential for good or evil.

In sum, Welsing offers no solution to the question of melanin and caucasoid behavior if melanin is the key factor. Reducing a person's negative conduct to biological influences leaves little hope for corrective behavior. Focusing on melanin as the causative factor in human behavior is futile because to do so merely limits the speculative possibilities not only of science but also of people, because the world is constantly changing. In relation to the conditions of the evolving environment in which people find themselves, melanin may be, perhaps, biologically advantageous or a handicap. The psycho-historical (Forbes, 1992; Bradley, 1991) rather than the psychogenetic model may better explain differences in behavior.

Michael Bradley (1991), a white Canadian, has also explored some of the causative factors of behavior. Specifically, he has uncovered the prehistoric sources of Western man's racism, sexism, and aggression by linking caucasoid behavior to the effects of the ice age. Surprisingly, Sigmund Freud addressed this "iceman inheritance" years earlier, but Freud's ideas were not popularized. Bradley reveals that Freud wrote an unpublished 12-page paper suggesting that Western man's psychosexual aggressions and ambivalences were caused by glacial evolution during the last European Ice Age.

For some reason, African-centered scholars such as Leonard Jeffries have been solely attacked for using the term "ice people." The critics seldom acknowledge the fact that many white intellectuals have written manuscripts and theses on this very postulate. The media has vilified Jeffries, though he in no way originated the theory.

Another author who has written a psychohistorical analysis is Jack Forbes, a Native American. Forbes wrote *Columbus and Other Cannibals* (1992), discussing the destructive anti-life behavior of Western societies. Forbes believes there is a sickness plaguing society, enabling exploitation, imperialism, and terrorism to thrive. He does not confine his argument to people of European descent, but believes that their psychohistorical experiences have contaminated other cultural groups in the same way a disease inhabits an organism:

> "For several thousand years human beings have suffered from a plague, a disease worse than leprosy, a sickness worse than malaria, a malady much more terrible than smallpox....Imperialism, colonialism, torture, enslavement, conquest, brutality, lying, cheat-

12

ing, secret police, greed, rape, terrorism—they are only words until we are touched by them. Then they are no longer words, but become a vicious reality which overwhelms, consumes and changes our lives forever...This is the disease, then, with which I hope to deal—the disease of aggression against other living things and, more precisely, the disease of the consuming of other creatures' lives and possessions...I call it, this disease, this wetiko (cannibal) psychosis, it is the greatest epidemic sickness to man." p. 9-10.

Neither Bradley nor Forbes discusses genetics or melanin in his argument. Using a psychohistorical approach, their analyses of Western man's behavior are more credible compared to an untestable hypothesis related to melanin. It is easier to research a psychohistorical model rather than a psychogenetic one.

For example, individual and group behavior can clearly develop out of specific environmental conditions. If an environmental factor, e.g., sunlight, is a biological threat, individuals living with that threat will cognitively develop ways to eliminate it. Given such a scenario, it is more plausible to posit that people use their cognitive skills to shape and mold their environment to cope effectively. The motivation and drive to control the world stems from a mental state that is survival-oriented. Any influence by a biological substance like melanin would be minor, or none at all.

In sum, the collective mentality of white people did not develop as a psychological defense to prevent their genetic annihilation resulting from an inability to produce melanin. However, we can state that the collective behavior of white people developed as a result of a threatening psychohistorical experience, i.e., a cold, harsh, resource-limited environment. Historically, nature was not a severe threat in this way to people of African descent. Thus, they had no need to respond aggressively to nature or to the world's inhabitants.

CAROL BARNES

Carol Barnes is an original member of the melanin research group that is called KM-WR Science Consortium. As a polymer chemist, Barnes has studied melanin for nearly two decades. He has played an integral role in helping African-centered scientists and theoreticians understand melanin. His knowledge of general and organic chemistry has added much information to the African perspective on melanin, and his knowledge of chemistry is where his strength is demonstrated.

Barnes began studying the properties of melanin in rubber compounds. From these experimental investigations, he has attempted to explain that internal and external melanin in the human body function similarly to the synthetic melanin he has studied in the lab. Judging from his laboratory experience, his expertise appears to lie in the technical area of organic chemistry.

Barnes' book, *Melanin: The Chemical Key to Black Greatness* (1988), is helpful for directing a person to scientific literature on melanin. Although the information is technical, his "Melanin Black Greatness" series has been written to remove the technical nature of the subject so the layperson can comprehend the discussion. He has researched the harmful effects of toxic drugs on melanin centers within the black human. Besides making black people "great," however, it appears as though the toxic effects produced by drugs on melanin centers in the brain can possible make black people "weak" or more vulnerable to certain types of drug addiction.

His book covers a wide variety of topics ranging from the structure and properties of melanin, its location in the body, and how toxic drugs affect or alter melanin functioning. He has provided 24 figures, 4 tables, numerous illustrations, 61 references and a helpful glossary. The scientific references are a good resource for the scientist interested in further reading; however, it was difficult to follow Barnes' interpretation of these articles and books.

Another problem with Barnes' book is the style. Distinctions between fact and opinion are vague, and such a style can mislead the reader and increases the confusion surrounding melanin research. The reader may not know whether a fact is proven and validated through scientific research, or whether it is merely a matter of conjecture.

Barnes suggests that "melanin's chemical structure also allows us to make proposals and predictions as to its potential capabilities." Emphasizing predictions and proposals without providing ideas for research can cause the reader, especially a novice, to misinterpret the author's intentions. It would be more beneficial to science as well as to current cultural debates to propose research paradigms that are actually feasible. An alternative presentation would be to clearly delineate the author's ideas as separate from scientific research, eliminating the confusion associated with many melanin presentations. Thus, the public could more safely distinguish between proven facts and personal belief.

The following statements are some of Barnes' predictions and proposals:

1) Melanin causes the expressive, flamboyant and cocky nature of the black human (toughness). p. 7.

2) Melanin granules are "Central Computers" and may analyze and initiate body responses and reactions without reporting to the brain. p. 39.

3) Melanin is responsible for the existence of civilization, philosophy, religion, truth, justice, and righteousness. p. 40.

4) Melanin is a civilizing chemical and acts as a sedative to help keep the black human calm, relaxed, caring, and civilized. p. 40.

These proposals and predictions are misleading and should not be made unless there is scientific research to clearly support such ideas. These unsubstantial statements are legitimate reasons for critics to say that melanin theorist are "teaching a strange doctrine related to black superiority." Furthermore, critics of African-centered researchers would highlight these types of ideas to imply that all African-centered melanin researchers are "racists" or pseudoscientists (Ortiz de Montellano, 1993; Jaroff, 1994).

Overall, Barnes' book can be used as a guide to existing scientific literature on melanin. He concisely explains some of the essential properties of melanin and the harmful effects related to toxic drugs. The novice as well as the scientifically advanced reader can utilize this document as a reference guide for literature pertaining to melanin.

In 1993, Barnes demonstrated his creative abilities by self-publishing another book titled, *Jazzy Melanin: A Novel*. The novel is an innovative approach to helping readers conceptualize melanin functioning. For readers who enjoy novels pertaining to science, the book should be enjoyable. However, if a researcher is seeking clarity in the area of melanin research, the novel is not recommended. A novel is a fictional prose narrative, and a serious topic such as melanin functioning should not be associated with make-believe tales. Although Barnes provides 10 excellent graph illustrations relating to melanin, the line is vague between actual facts and Barnes' "jazzy" approach to the presentation of melanin which Barnes espouses.

Barnes' specialty is chemistry, and readers could benefit more from his knowledge if he kept his focus on the technical aspects of melanin chemistry. His intellectual trek into disciplines outside chemistry, however well-intentioned, fails to guide readers to a clearer understanding of melanin functioning. Based on my discussions with people who only vaguely understand melanin and other related concepts, I can only conclude that the existing literature from a black perspective is quite confusing to many. Unfortunately, Barnes' recent publication does not resolve this confusion or offer us important new knowledge about melanin. Although Barnes is beyond trying to convince us that melanin is "the chemical key to black greatness," communicating actual facts about melanin should remain a top priority.

In conclusion, the novel should not be recognized as a major source of information pertaining to melanin functioning insofar as human behavior is concerned. Brilliant scientists such as Barnes ought not demean their technical skills by writing half-serious books. Presently, the severe controversy in which melanin is enmeshed presents us with a need for new and highly professional melanin studies.

MALACHI ANDREWS

Malachi Andrews is a kinesiologist based in California.* As a kinesiologist, he specializes in the study of movement. He has worked in the area of physical education, and he is one of the founding members of KM-WR Science Consortium. Andrews has one of the most vibrant personalities of all African-centered scholars. His writing style in his 79 page book, *Color Me Right...Then Frame Me in Motion* (1989), reflects his personality. He moves from topic to topic, and it may be difficult to understand melanin functioning if this book is your only resource. Similar to Welsing's work, there was no major scientific evidence reported by Andrews to support his views on melanin.

Andrews writes and speaks in strong African-centered terminology, and the content of the book is a mixture of essays, poems, personal experiences, quizzes and test questions, illustrations and definitions. The author does not emphasize the science of melanin. Once again, this quasi professional kind of literature can mislead the reader. Furthermore, the incomplete bibliography denies readers the opportunity to pursue the subject more deeply or to examine his source material for themselves.

RICHARD KING

Richard King is a practicing psychiatrist in North Carolina and the author of *The African Origin of Biological Psychiatry* (1990). He has a lot in common with the previously mentioned authors in that he is a psychiatrist like Welsing and a co-founder and member of KM-WR Science Consortium along with Barnes and Andrews. He has served as president of KM-WR, and he gives numerous lectures throughout the country. Since he has extensively studied Ancient Egyptian history, he is able to synthesize ancient and modern science. Similar to Welsing, his book is a compilation of his essays that he previously wrote in the *Journal of Uraeus*. From these writings, King is most known for his Black Dot...Black Seed concept for what is commonly known as the collective unconscious.

King links ancient wisdom and modern research to inform us that early African scientists knew much about brain functioning and consciousness. If you would like to know about the symbolism of Ancient Egyptian culture and how significant this information is to modern research, this is the book to read.

King makes reference to blackness throughout his text. Blackness, according to King, relates to "...the black seed of all humanity, archetype of humanity, the hidden doorway to the collective unconscious-darkness, the shadow, primeval ocean, chaos, the womb, doorway of life." Furthermore, "the chemical key to life and the brain itself was found to be centered around black neuromelanin." It is in Chapter two of his book that King focuses on melanin functioning. Other than Chapter two, there is minimal discussion on melanin as a major topic.

He has done a remarkable job of researching ancient African literature and uncovering modern scientific evidence to support his views. Most of the articles he reviewed are from the 1930s to the 1970s to support the fact that early African scientists were well informed about melanin and our complex nervous system. For example, there is an upper brainstem region known as the locus coeruleus that secretes the neurotransmitter norepinephrine. Norepinephrine is primarily released from this brain region for global activation of the left and right hemispheres and subcortical structures. The locus coeruleus consists of several dark melinated nerve cells, and according to King, the Indian Sanskrit word coeruleus is derived from the Ancient Ethiopian name, Celeno, which means black. The locus coeruleus is the focus of King's Black Dot...Black Seed concept.

King relates past and present research by stating, "as Western science investigates the locus coeruleus, increasingly more biological evidence emerges that directly supports the ancient African concept that coeruleus is a Black Dot doorway to the collective unconscious." His scientific research associates the role of the locus coeruleus as a doorway to the collective unconscious with the fact that the locus coeruleus is the uppermost point in the melinated nerve tract that extends from the brainstem to the midbrain. According to King, this neuromelanin or "Amenta" nerve tract is the black internal core in all humans and is the "doorway that opens into an all black hall of blackness...," the doorway to the collective unconscious.

There is much more to King's thoughts and ideas, but the Black Dot...Black Seed is the central theme relating to melanin. King's book is recommended for any reader who has an interest in African history and science. In terms of criticism, a few points can be made. Since the black community considers King one of the "experts" on melanin, it is disappointing that he does not thoroughly cover the processes and mechanisms by which melanin functions. He does state, however, that he is only a humble "student" in this complex area of study. Also, his book is a compilation of his earlier writings, which explains the minimal use of current research articles.

In the latter part of 1993, King compiled a notebook of material that was mostly related to ancient African History. The notebook is titled, *The Black Light: The Face of Ra.* Prior to 1994, his document was organized in a three ring binder notebook not prepared for mass distribution. In 1994, he organized this notebook and other writings into a document called, *Melanin: The Key to Freedom.* The documents contain historical and scientific material that King has collected since publishing *The African Origin of Biological Psychiatry.* Neither document is an exhaustive presentation on melanin research. However, numerous up-to-date research articles are referenced, some of which are reviewed further along in *The Science of Melanin.*

As we conclude this critical analysis of King's concept of the Black Dot...Black Seed, we prepare to move on to the next chapter - The Biology of Blackness. We will now embark on another African adventure into the darkness of melanin. Out of the darkness of melanin we will see the shining light of optimal health.

*Author's note: Sadly, Malachi Andrews died in 1994 after the completion of this book.

2

The Biology of Blackness

The dark color or pigment of the human body that is found internally as well as externally is called melanin. External melanin can be found in body regions such as hair, skin and eyes. Internal melanin is located in body regions such as the central nervous system, heart, liver and various other internal organs. Although external and internal melanin are found in different body sites, both are embryologically derived from similar cells. To better appreciate the relationship between external and internal melanin, we will need to discuss developmental embryology.

No matter what topic is being investigated, it is necessary to start from the beginning in order to comprehend sufficiently the analysis that is being studied. Developmental embryology is one discipline that deals with the origins of human life. We can utilize this area of science to help us better understand melanin's role in the early stages of human development. We will only present pertinent information in this area; however, an introductory textbook on developmental embryology is suggested for further reading.

DEVELOPMENTAL EMBRYOLOGY

The wonder of human reproduction begins with the uniting of two specialized sex cells called the sperm and the egg. Sperm cells and egg cells are called gametes when they are separated. However, the term zygote is used when they are united to form the human organism (embryo). From the earliest stages of human development, melanin is found in several critical sites in the embryo.

After fertilization, the zygote will transform through several stages of cell replication. The cells rapidly divide during the early stages of development to form three distinct cell layers. These layers are the endoderm, mesoderm and the ectoderm, and they are collectively known as the primary germ layers (we are specifically interested in the ectoderm). The endoderm and the mesoderm are less important for the purpose of our discussion. However, the derivatives of both layers will be mentioned.

The prefix to these primary germ layers gives a clue to the various body parts that will be derived. For example, the inner embryonic layer, the endoderm, forms the digestive system and the respiratory system (Oppenheimer and Lefevre, 1984). The pancreas, liver and thyroid gland are also derived from this innermost germ layer.

The mesoderm, middle embryonic layer, is the most intricate of the three layers because of the numerous subdivisions. For example, the major divisions of the mesoderm are divided into the epimere, the mesomere and the hypomere. These are the derivatives from the subdivisions of the mesoderm:

Epimere   -   Myotome - back muscles.
                Dermatome - dermis of the skin.
                Scleratome - vertebral column.

Mesomere -   kidney, gonads and associated structures.

21

Hypomere -   lining of the body cavity, smooth gut muscle, heart, blood vessels and associated structures.

To understand the important role of melanin during the early stages of embryonic development, we are interested in the derivatives of the outer embryonic layer--the ectoderm. The ectoderm is composed of three regions: the prospective neural tube; the prospective neural crest; and the prospective epidermis. It is within these three regions that melanin plays its first key role in maintaining life. In the midst of this embryological darkness, an explosion (Big Bang) will occur that will transform the tiny mass of cells into a complex human being similar to a minature solar system.

Each region of the ectoderm is further differentiated into specialized body parts. The *neural tube* is differentiated into a) the brain, b) the posterior pituitary gland, c) the optic vesicles, d) the spinal cord, and e) the motor nerves that originate in the ventral portion of the neural tube and innervate muscles.

The *neural crest* derivatives consist of cells that migrate to distant parts of the body. These migrating cells form sensory nerves and ganglia, which receive impulses from the following sites: 1) sense organs; 2) autonomic ganglia; 3) the adrenal medulla; 4) all of the pigmented retina cells, which are derived from the neural tube; 5) the cartilages in the voice box and head; and 6) some of the ectodermal muscles.

The *epidermal layer* can be divided into cells derived from epidermal thickenings and those derived from the rest of the epidermis. The thick epidermal derivatives include some of the cranial nerves, the lens of the eye, the olfactory structures, the inner ear, and the taste buds. The remainder of the epidermis forms the following structures: the outer layer of the skin; the hair and nails; the linings of the mouth and anus; and the anterior pituitary.

Developmental embryology is much more detailed than what is discussed here. This concise summary is provided to inform the reader that melanin is not just a colored substance, but an organizing molecule (Barr, 1983) present at the early stages of human growth. Nature has provided the human body with an efficient network of biological systems, and melanin plays a critical role in ensuring that these systems are properly formed and maintained.

It is important that one comprehend melanin's role in the various structures that were mentioned. In these body regions, melanin provides many functions such as facilitating the conversion of energy and protecting cells from toxic substances. Furthermore, one should attempt to understand the relationship between melanin and the migration of cells to specific destination sites such as the nervous system and skin. Before introducing more scientific terms that should assist in better understanding the various roles of melanin, we will next expound on the physiological origin of blackness.

THE PHYSIOLOGICAL ORIGIN OF BLACKNESS

In scientific terminology, the suffix *cyte* is used to describe a specific type of cell. For example, the sex cells that divide to reproduce other sex cells are called spermatocytes for sperm cells and oocytes for egg cells. A similar biological description is used for those cells that produce melanin. In other words, the cells that divide and produce melanin are called melanocytes. Melanocytes are pigment-forming cells that are mostly derived from the neural crest. Therefore, the orign of blackness or pigmentation is a result of melanocyte functioning.

From the beginning of life, melanin is located in a strategic location (neural crest) to migrate eventually to numerous regions within and outside of the body. Psychological processes such as sensation and perception will be dependent upon melanin functioning for heightened physiological arousal and extrasensory perception. In other words, the properties of melanin (see Chapter three) can optimize the functioning of our physiological systems, and its absence during fetal and/or advanced human development can be life-threatening. If melanin was not important for the early stages of human development, then why is it found in the life form during critical stages of human development? One would need to ask the question, "What would be the repercussions if melanin were absent from the developmental process?"

23

It is not by accident that our existence begins in darkness (blackness) in the zygote and in the womb. Interestingly, there are other ways to express the importance or significance of blackness at the early stages of development. For those that read the Bible, the clearest metaphor can be found in the first four verses of *Genesis*.

1. *In the beginning God created the heaven and the earth.*
2. *And the earth was without form, and void; and darkness was upon the face of the deep. And the spirit of God moved upon the face of the waters.*
3. *And God said let there be light: and there was light.*
4. *And God saw the light, that it was good: and God divided the light from the darkness.*

For those who do not find much credence in Biblical teachings, another metaphor can be described. For example, the creation of thoughts and ideas begin in an unformed state in our minds. The thoughts and ideas that come from the mind are found in the recesses of the subconscious (dark) mind. Only when your thoughts and ideas are brought into fruition and turned into action (light) can we say that something will be formed. Therefore, our thoughts and ideas first occur in the blackness of our imagination to later form into a conscious manifestation.

To conclude this chapter, melanocytes are cells that produce melanin, and they are derived from the neural crest. Since melanin is associated with the distribution of numerous types of cells to other destination sites in the body, it is apparent that there is a critical role for the darkness provided by melanin. From the depth of the developing zygote to the vastness of the dark universe, melanin may be the link between the material and spiritual realms of existence (see Chapter ten).

On the cellular level, melanin is a chemical secreted by specialized cells called melanocytes. On a supracellular level, the skin is the largest organ in the body and it contains melanocytes. It is my belief that melanin plays a role as a hormone secreted by the skin. By contrast, melanin in the brain (neuromelanin) seems to be very stable and is not normally secreted from the cells where it is formed (Lindquist, Larsson and Lyden-Sokolowski, 1987). In sum, melanin has additional roles in addition to tanning the skin, and these revelations concerning melanin functioning will be further discussed in the next chapter.

3

Advantages of
Melanin Functioning

This, of all the chapters, is most important in understanding the
special properties of melanin in living organisms. The information in this chapter will be critical for dispelling all myths and
skepticism associated with melanin's functional capabilities. If you
have general knowledge of chemistry, biology, physiology and neuroscience, then this chapter should be very informative. If you are a
nonscientist, then accept the challenge and "endarken" yourself about
the concepts that make melanin such an important biological substance.

If you doubt that melanin has a functional role in life, then study
this chapter. If you think melanin research is "mythological fantasiz-
ing," then study this chapter. If you reluctantly discuss melanin in
your lectures or presentations, then study this chapter. Answers to
most of the basic questions concerning what melanin can do to affect
the living organism will be analyzed. As you read the other chapters,

it is suggested that you refer back to Chapter three for scientific evidence to support statements made about melanin functioning in various regions of the human body.

These introductory statements are provided because this chapter will probably have the most impact on how the general public views melanin. Although the discovery of melanin and its location in the skin have been studied for decades, many people are not aware of melanin research. This book does not intend to be the first or last book on melanin. However, it is written, with this chapter in particular, to review some research that clearly suggests a functional role for melanin beyond skin pigmentation.

The function of melanin in the human body can be analogous to chlorophyl in plants. Chlorophyl is not only the pigment located in plants, but it is also utilized in the very important process of photosynthesis. Without chlorophyl, there would be impaired conversion of carbon dioxide and water into oxygen and glucose. Similarly, melanin is not only the pigment on the external surface of the body, but it is also found internally to assist in maintaining a healthy organism.

Early research presupposed that melanin was a waste product that had no physiological function (Graham, 1979). Since melanin was thought to be the end product in catecholamine synthesis, it was generally considered to be a principally inert waste product from the synthesis and turnover of dopamine that occurs in pigmented brain cells (Bazelon, Fenichel and Randall, 1967; Bogerts, 1981). Other researchers have demonstrated that melanin in the skin and melanin in the brain form differently (Rodgers and Curzon, 1975). However, both are critical for proper cell functioning.

Some of the physiological roles of melanin include its protective capabilities, the fact that it can act as a deposit site for free radicals (highly reactive chemical substances), and its ability to affect nerve impulses. Along with these roles, we will discuss melanin's ability to act as a semiconductor to transform energy. Essentially, the bioelectronic properties of melanin are what advance the functioning of all living organisms that have melanin in various regions of the body.

27

PROTECTIVE CAPABILITIES AS A NEUTRALIZER

Melanin has protective capabilities in every region of the human body where it is located. Externally, it protects the skin from the harsh elements of mother nature. For example, melinated skin provides protection from the harmful effects of ultraviolet radiation and dangerous chemicals, and it helps to retard the aging process. The youthful and lustrious appearance of the skin in people who are highly melanized is due to the cumulative effects associated with melanin functioning in the skin.

Other noticeable indications that melanin is providing a protective role can be observed during the healing of wounds. When cells are damaged (e.g., a cut, tear or rupture), the appearance around the damaged cells turns dark in color. The dark color is an indication that melanin is preventing further cellular destruction, and it is stimulating the healing process. A similar effect is noticed when cut fruit is exposed to air.

Melanin is effective as a device for radiation-less conversion of the energy of harmfully excited molecules into innocuos vibrational energy (McGinness and Proctor, 1973). This conversion may deactivate metabolically excited molecules that can become potentially damaging to cells. Since there are no mechanical devices involved, that is why this conversion is called radiation-less. The important point is that melanin helps to ensure that the spread of further cellular damage is neutralized.

Melanin is also known to protect against dangerous free radicals. Free radicals are highly reactive chemical species that have an odd number of electrons, and hence, one unpaired electron. It has been proposed (Commoner, Townsend and Pake, 1954) that melanin acts like a deposit site or sink for unpaired electrons, thus removing reactive free radicals. Peroxides are examples of chemical substances that can lose an electron and change into dangerous and cytotoxic substances.

The stable nature of melanin also gives it the unique feature to accumulate several biological compounds. For example, various amines, drugs, and a number of metals can be retained in melinated

28

regions of the body for a very long time (Larsson and Tjalve, 1979). Lindquist, Larsson and Lyden-Sokolowski (1987) suggested that melanin's capacity to accumulate and bind a variety of compounds can allow neuromelanin to protect the cells that harbour the pigment by keeping potentially harmful substances bound, and thereafter, slowly releasing the agents in low, non-toxic concentrations.

What this means is that the level of melanin in the human body can significantly influence potential drug effects. As expressed by Barnes (1988), the slow release may cause certain drugs (e.g., cocaine) to have greater potency in highly melinated individuals. Therefore, drug addiction may be more severe in the black community (Barnes, 1988). When the skin is exposed to these biological substances, the damage can be observed by inspecting the skin. Within the brain, however, it is impossible to see the cellular changes caused by cytotoxicity. The brain is more delicate and complex than most other organs in the body. If the brain is saturated and overburdened with these cytotoxic molecules, then behavior is ultimately disturbed.

A third source of protection is its redox capacity. Besides cytotoxic molecules and free radicals, melanin may act as an electron-transfer agent for various chemical substances. In this capacity, melanin can protect cells and tissues against reducing or oxidizing conditions (van Woert, 1968; Gan, 1976; 1977). In other words, melanin is acting like a "chemical laboratory" where ever it is found. By assisting in the transferance of electrons, it ensures the safe conversion of potentially volatile chemical reactions.

In sum, all of these protective capabilities are to maintain a well-conditioned state of health. Melanin can protect against the numerous chemical and biological agents that exist in our environment, and subsequently, eliminate poor health. However, there is always the possibility that melanin might become overprotective. In the case of some neurodegenerative disorders (e.g., Parkinson's Disease), melanin may absorb too much and create a toxic situation (Calne, 1991). As an example, long term exposure to a toxic compound with high neuromelanin affinity may ultimately cause lesions in the cell (Lindquist et al., 1987).

The next role of melanin we will discuss is its effects on nerve impulses. We will discuss the nervous system at length in the next chapter. Therefore, we will not elaborate on the structure and function of the nervous system at this point.

NERVE CONDUCTION FACILITATOR

Communication in the nervous system is dependent upon electrical and chemical events. This electrochemical communication stimulates the transmission of signals from one end of the nerve cell and from cell to cell. The signal is transmitted in the form of a nerve impulse, and the bioelectronic properties of melanin can facilitate the conduction of nerve impulses.

Let us first look at the similarity between the nervous system and man-made physical structures that conduct electricity. Both electrical systems require a voltage, and the voltage determines the power that will emanate from the electrical source. The nervous system has electrical wires just like a simple household appliance or a complex computer. In the nervous system, a nerve impulse will only be generated when a specific voltage is reached. Once it is reached, activity in the nervous system is elicited.

You cannot plug in the body like an electrical appliance, but it does get electricity from another source--your diet. For example, we eat electricity in the form of electrolytes (charged particles, or ions) when we ingest nutrients. The electrolytes can form an electric current in our nervous system. If you can visualize the nervous system as a collection of wires soaking in water, then you can imagine how a current could be flowing through the body.

Water is electrically neutral, however, and water alone does not conduct electrical current. One scientist, Noboru Muramoto (1988) explained how ions dissolved in water can conduct an electrical current:

> "Common table salt is also electrically neutral, but when dissolved in the water, sodium ions ($Na+$) and chlorine ions ($Cl-$) separate and can conduct electrical current. When a substance is dissolved in water and

breaks down into ions, these are called electrolytes. Among the most important electrolytes in our bodies are hydrogen (H+), sodium (Na+), potassium (K+), magnesium (Mg+), calcium (CA++), chlorine (CL-), sulfur (S-), phosphorus (P-), oxygen (O-), iodine (I-), the hydroxyl ion (OH-), bicarbonate (HCO3-), and carbonate (CO3-)." p. 241.

After ingesting ions, these electrolytes set up a weak biological current in the nervous system that ranges from 40 to 120 millivolts. McGinness, Cory and Proctor (1974) performed some laboratory investigations on the electronic properties of melanin to reveal that the nervous system utilizes certain voltages, and that this electrical energy is harnessed by melanin. The authors stated that threshold switching at low gradients poses some theoretical questions. For example, melanin may act as an amplifier to "pump up the volume" on this low voltage. In effect, there is more activity elicited from melinated brain regions when compared to nonmelinated brain regions.

It is suggested that the bioelectronic properties associated with melanin can help to facilitate nerve conduction in the following three ways: 1) it can speed the pace of nerve impulses; 2) it can concentrate ions for high voltage generating activity; and 3) it can provide an electrochemical surge.

The first point indicates that melanin can help to transfer specific chemical molecules, and subsequently, speed the nerve impulse in highly melinated brain regions. Since melanin can act as an electron-transfer agent, it can cause the nerve cell to exhibit more activity by facilitating the electrical and chemical transmission of nerve impulses.

The second point relates directly to the accumulation of electrolytes or charged ions in melinated brain regions. The more charges in a cell can translate into a higher voltage in the cell. Therefore, melinated brain regions can aggregate charged ions and produce high voltage generating activity. As a result, the high voltage generating activity could facilitate the nerve impulse.

The third point suggests that melanin can cause a greater release of neurochemicals from nerve cells. Neural transmission requires the stimulation of certain ions or electrolytes across the cell membrane. As a result of this change in the cell membrane, neurochemicals are released from the cell to transmit a nerve impulse. Interestingly, there are several brain regions (e.g., substantia nigra, locus coeruleus, raphe nucleus) that have both high melanin content and neurochemical activity (e.g., dopamine, norepinephrine, serotonin). Therefore, melanin can increase the voltage, cause an electrochemical surge, and positively influence the release of neurochemicals.

It was first suggested by McGinness in 1972 that melanin may act as an amorphous semiconductor, that is a threshold switch. McGinness theorized about the electronic properties of melanin reported by Blois (1969; 1971), who observed a rise in the conductivity of melanin under an applied voltage. McGinness suggested that this rise might be a result of increased kinetic energy of the electrons leading to higher mobility and promotion to excited states. In other words, melanin may act as a threshold switch to "pump up the volume" from a lower voltage to a higher voltage.

In sum, the bioelectronic properties of melanin facilitate nerve conduction and generate much more activity than non-melinated brain regions. It is very critical that melanin is found in the nervous system because it helps to heighten mental awareness, it speeds reaction time, and it greatly enhances the capacity of the brain to transmit neural impulses. Consequently, the special properties associated with melanin in the nervous system can help to increase the survival of the organism.

ENERGY TRANSFORMER

The consistent appearance of melanin in living organisms at locations where energy conversion or charge transfer occurs (skin, retina, midbrain, and inner ear) is of particular interest.

Melanin is strategically placed in these locations to absorb and convert various forms of electromagnetic energy into energy states that can be used by the nervous system. Melanin is found in the skin to absorb ultraviolet light, in the retina to increase visual acuity and reaction time, in the midbrain to perform complex motor tasks like reverse dunking a basketball, and in the inner ear to amplify the thumping bass in music and/or to maintain rhythm and equilibrium when rapping to music. In all of these instances, melanin may be acting as an electronic device (McGinness et al., 1974).

The fact that melanin is black or dark in color could explain how it is functioning as a converter of energy. Since dark skin or any black substance absorbs heat, light is not reradiated, but is converted to rotational and vibrational degrees of freedom (McGinness and Proctor, 1973). Contrary to blackness, whiteness reflects light. As a result, pigmented cells are more capable of converting energy versus nonpigmented cells. Although there are comparable distributions of neuromelanin in the brain, darker colored people have a more modified external pigmentary system that has a greater capacity for charge transfer.

There are other physical mechanisms by which melanin can influence electrical energy. It has been proposed (Lacy, 1984; McGinness et al., 1973) that phonon-electron coupling is the possible transducing mechanism in bioelectronic processes involving neuromelanin. In other words, melanin is involved in the transformation of one form of energy into another by coupling together two different physical states (vibrations and electricity). The black appearance of melanin suggests that the phonon-electron coupling in this material may be particularly efficient for transforming energy. In the case of sunlight, the absorbed radiation is retained and transferred to excited electronic states. This excited physical state can explain the pleasant mood one may have on a bright, sunny day.

At the level of the brain cell (neuron), however, there are significant neurophysiological functions that are constantly improving the excitation and conductivity of the nervous system. From that point, changes in the electronic nature of neuromelanin could generate vibrational energy (phonons) capable of affecting nerve impulses. Stimulation in the nervous system occurs by way of an action potential. Lacy (1984) proposed a process involving action potentials and electron-phonon coupling:

33

1. Changes in neuromelanin semiconduction, perhaps involving threshold switching, would cause transient effects on phonon production by neuromelanin.

2. Phonons produced by neuromelanin granules are transmitted through the cell and absorbed by membrane molecules.

3. Vibrational interaction between membrane molecules would influence polarization of the membrane and subsequently affect membrane permeability to ions.

In sum, the membrane is a very important cellular structure that is critical for ions to flow in and out of cells. The proper flow of ions through the membrane can generate action potentials and elicit normal brain functioning. The electrical properties of neuromelanin can enhance the performance of the brain's electrical activity by acting as a semiconductor. Because membrane permeability is essential to nerve conduction, semiconduction by neuromelanin could potentially increase the rate of action potentials.

Semiconduction and Superconductivity are major areas of study in physics. Scientists who study these advanced physical processes are developing better and more efficient ways to harness energy. Research on melanin's semiconductive capabilities is very important to modern technology because the naturally occurring physical processes of melanin are what scientists would like to replicate.

Melanins respond at a critical applied field by changing their conductivity (Filatovs, McGinness and Cory, 1976). The nature of the response depends on the physical environment (i.e., in vivo or in vitro) and, the response can be classified into two categories: 1) Threshold switching; and 2) Memory switching.

Threshold switching occurs when a sample cycles from an off (low conductivity) to an on (high conductivity) state at a critical electric field and returns to the off state when the electric field is removed. Memory switching, on the other hand, refers to a sample which remains in the on state when the field is removed but can be restored to the off state by larger electric fields or currents. Both threshold and pseudomemory switching have been reported in melanins (McGinness, Corry and Proctor, 1974; Culp, Eckels and Sidles, 1975).

These switching mechanisms are very significant findings. In fact, these physical mechanisms are probably the most important scientific evidence to indicate why neuromelanin is advantageous to the living organism. In the study conducted by Filatovs et al. (1976), they stated that melanins are capable of absorbing other molecules and expressing this interaction by a conductivity change of as much as ten orders of magnitude. In essence, melanin is "pumping up the volume" by absorbing, storing and elevating neural impulses. The scientific evidence reveals that when it is necessary for a person to transcend a mental obstacle and overcome adversity, neuromelanin can switch to a higher stage of functioning and engage the brain for improvisation.

How do we best capsulize these findings? First of all, we need to understand melanin's abilities to transform one form of energy to another (electron-phonon coupling). By transforming energy, melanin can maximize input into the body's command center—the nervous system.

Secondly, we should realize melanin's capabilities to store energy. This storage might be called memory. We can use the word memory since melanin can make use of energy input that was previously used. The memory for previously used neural information can maximize input into the brain. For scientific evidence concerning melanin's storage capabilites, McGinness (1985) has summarized the effects as follows:

1. When a current is passed through melanin, a pH gradient is formed and energy is stored.

2. If a pH gradient is developed across melanin, it will discharge a current.

3. Drugs which bind to melanin modulate the stored
charge.

In sum, neuromelanin may affect neurons in their vicinity by
storing energy and using it for future neural transmission. Depending
on the brain region, the spread of stored energy in certain locations of
the brain can enhance mental abilities that are associated with that
particular brain region. Moreover, drugs can alter the normal storage
functions provided by melanin and positively or negatively affect
behavior.

Let us further look at what could happen around pigmented nerve
cells. Consider a neuron which has not been fired. It will be in
equilibrium with the neuromelanin in its immediate surrounding.
Once the neuron has fired, the neuromelanin will respond to the
changing electrical and ionic gradients. If the neuron recovers before
the neuromelanin, then the conditions required for production of an
action potential may be changed to potentially influence the firing
frequency. If the time it takes for the neuromelanin to return to
equilibrium with the neuron is extended to the time in which the
neuron is stimulated to fire again, then the neuromelanin would
contain a short time memory of the previous firing event. According
to McGinness (1985), this could allow pigmented neurons to process
information differently than their nonpigmented counterparts.

The frequency of nerve impulses is a very important mechanism
for providing activity in the nervous system. According to Feldman
and Quenzer (1984), the frequency of nerve impulses is an important
consideration of neural function because it plays a significant role in
information transmission and in the transmission of nerve impulses
from one neuron to another. Since the voltage must reach a certain
point to engage the neuron for action (all-or-none principle), information about the intensity of a stimulus cannot and is not forwarded to
the brain by amplitude modulation (AM). Remarkably, information
about intensity differences between stimuli is transmitted by a frequency modulating (FM) system (Feldman and Quenzer, 1984). The
bioelectronic properties of neuromelanin help to facilitate this process
because the greater the stimulus intensity, the higher the nerve

impulse frequency. As stated earlier, melanin can "pump up the volume" when transmitting neural messages.

In conclusion, we have reviewed numerous research articles and several theoretical viewpoints to emphasize the advantages of melanin functioning. We have in no way suggested that melanin can make one "superior" to those who lack melanin. It is a hollow argument to speak of superiority because people who lack skin melanin could have a biological advantage under specific conditions where the role of skin melanin is not as important.

However, the absence of melanin in the brain (neuromelanin), can be deleterious. For example, the deterioration of melinated brain structures is associated with severe disturbances in behavior such as Parkinson's disease, phenylketonuria and mental retardation. The bioelectronic properties of melanin are what essentially advance the functioning of the nervous system, and the information in this chapter was presented to emphasize that melanin is essential for maximizing human physiology. In Part Two, we will present information on several physiological systems that are directly and indirectly influenced by melanin.

Part Two - Melanin Physiology

Physiology is the branch of biology dealing with functions and vital processes of living organs. Human physiology attempts to explain the specific characteristic and mechanisms of the human body that make it a living being (Guyton, 1982). Guyton has explained that the human being is actually an automation. The fact that we are sensing, feeling and knowledgeable beings is part of this automatic sequence of life. In Part Two, we will try to explain how the body is dependent upon melanin for automaticity and for functioning at an optimal level.

4

Nervous System

As stated in Chapter Two, the neural crest melanocyte precursors migrate from the neural crest to a variety of sites in the body. The most important influence appears to be melanin's role in providing the framework for the body's command center--the nervous system. The nervous system maintains control by integrating all bodily functions. Our awareness of the environment is made possible by the integrated functioning of a group of tissues composed of highly specialized cells possessing the characteristics of excitability and conductivity. Melanin can enhance both excitability and conductivity in the nervous system to give an optimal level of performance. In this chapter, we will discuss melanin's role in enhancing the functioning of the nervous system.

Let us begin by discussing the organization of the nervous system. The first two divisions of the nervous system are the central nervous system (CNS) and the peripheral nervous system (PNS). The CNS includes all parts of the nervous system in the bones of the skull or spine. The brain and spinal cord, therefore, are the major components of the CNS. Nerves that come into and out of the CNS make up the PNS. The PNS includes 12 pairs of cranial nerves and their branches and 31 pairs of spinal nerves and their branches.

The smallest and basic unit of this complex network of tissue is the neuron. The neuron, or nerve cell, consists of a cell body and two types of processes called dendrites and axons. Dendrites are branch-like processes that carry impulses toward the cell body. Axons are single, elongated cytoplasmic extensions carrying nerve impulses away from the cell body. One of the key roles of melanin is to facilitate the conduction of nerve impulses through certain regions of the nervous system.

Early reports characterized the brown-black intracytoplasmic granules found in CNS neurons as "neuromelanin" (Lillie, 1955; 1957). Special histochemical techniques have distinguished the neuromelanin from external or skin melanin in humans. In the late 1960s, Bazelon and colleagues reviewed studies done on neuromelanin in the human adult and infant nervous system (Bazelon, Fenichel and Randall, 1967; Fenichel and Bazelon, 1968). These authors reported that the brainstem provided the most pigmented cells. According to Olszewski (1964), 12 nuclear masses containing black pigmented cells have been systematically mapped in brainstem nuclei.

Bazelon et al. (1967) noted the striking parallel between the anatomical maps of neuromelanin neurons and catecholamine cells in the brainstem. This is a very important point because neuromelanin may enhance the functioning of certain neuro-chemicals in the brain. For example, the locus coeruleus (LC), which secretes norepineph-rine, is black because it contains large amounts of melanin. Amaral and Sinnamon (1977) suggested that LC melanin is the same as skin melanin. Other researchers, however, state that the chemical compo-sition of brain melanin is known to differ in many respects from other melanins (Moses, Ganote, Beaver and Schuffmann, 1966; Barden, 1969).

From an African-centered perspective, King (1990) has stated that the LC is an anatomical brain site that is important for linking the human mind to the spiritual world. King has thoroughly studied ancient African science, and he believes that the LC is biological evidence to support the ancient African concept that the LC is a Black-Dot doorway to the collective unconscious. According to King, the 12 black pigmented brain sites reported by Olszewski (1964) make up the "Amenta" nerve tract. Furthermore, King suggests that the neuromelanin of the "Amenta" nerve tract is a spiritual doorway to the unconscious mind.

Structurally and functionally, the LC provides the principle nerve supply of norepinephrine to many areas of the brain such as the cerebral cortex, hippocampus, cingulate gyrus and amygdala. These brain sites make up the circuitry for the limbic system, which is primarily responsible for controlling emotional behavior. In addition to emotions, the hippocampus has a key role in memory formation. Also, norepinephrine has been associated with global activation of the brain for alertness.

Therefore, emotions, memory, and alertness are some mental processes generated by the brain that are highly dependent on the proper functioning of LC norepinephrine.

Another black pigmented brain region that is critical to proper brain functioning is the substantia nigra. The name substantia nigra literally means black substance, and it is neuromelanin that makes it black in color. Similar to the LC, the substantia nigra has a parallel between the anatomic distribution of melanin and catecholamine containing cells. However, the substantia nigra is a midbrain nucleus that contains dopamine producing cells rather than norepinephrine. Fenichel and Bazelon (1968) suggested that neuromelanin may be a marker of active catecholamine metabolism to explain the parallel distributions. In addition to revealing active catecholamine metabolism, melanin can also attract potentially dangerous environmental agents.

The fact that numerous environmental agents could be toxic to limited neuronal populations can be exhibited in the neurodegenerative disorder known as Parkinson's disease. Parkinson's disease is pathologically characterized by destruction of the dopaminergic cells in the substantia nigra. There have also been reports of degenerating cells

43

in the ventral tegmental area and the LC (Kaplan and Sadock, 1988). If neuromelanin in the substantia nigra deteriorates, then there will be a loss of dopamine production. The major clinical manifestation of such a problem would be tremor, rigidity, loss of coordinated movement, and disturbances of posture. The disease occcurs usually between the ages of 40 and 60 years of age. Current research, however, reveals that Parkinson's disease can manifest at a much younger age.

The exact cause of Parkinson's disease is unknown, but there have been links to environmental agents. For example, some young people in California who experimented with street drugs ingested a synthetic heroin known as MPTP. The metabolite of MPTP, MPP+, caused these young people to develop Parkinson-like symptoms. Ostensibly, melanin's ability to accumulate free radicals and metallic substances can destroy the functioning of pigmented brain cells. The specific role of neuromelanin is narrowly defined in the scientific literature; however, research on neurodegenerative disorders such as Parkinson's disease can provide critical information.

Another issue in melanin research that deserves further attention is the formation of melanin in the brain. In 1975, Rodgers and Curzon performed a quantitative radiometric assay to measure melanin formation in the human brain in vitro. They reported that melanin formation was detected in all brain regions studied and was highest in substantia nigra and striatum. Fenichel and Bazelon (1968) reported that melanin was not consistently observed at birth. The accumulation began during the first five years of life and increased steadily throughout childhood. It was not reported in the study, but it would be interesting to know something about the ethnicity of the infant brains. There is a distinct possibility that blacks and whites differ in melanin formation during infancy. Critical evidence for a difference can be found in several developmental studies that investigate infant psychomotor abilities. (See Chapter 9 - Melanin and Early Childhood Development.)

Since neuromelanin is found deep within the brain where no sunlight reaches, we cannot attribute the dark pigment to the effects of sunlight. Genetics is one area of study that can help us find answers to why dark-pigmented cells are found where there is no light. For example, Schneider, Shelton and Kuff (1975) reported a significant finding in animals that associated DNA with melanin granules. Melanin granules were isolated from the Cloudman S91 mouse melanoma and from Amphiuma liver in highly purified form. Their results suggested that DNA may be a component of melanin granules, and the component may differ in both tumors and normal tissues. In other words, all types of melanins are probably distinguished at the level of DNA to support the viewpoint that neuromelanin is species-specific (Mann and Yates, 1983; Marsden, 1983). In humans, levels of neuromelanin may vary depending upon an individual's ethnicity. Accordingly, variation in neuromelanin could be genetically programmed.

Additional support for the above finding is the fact that melanoma contains granules varying in their degree of melanization. Schneider et al. stated that it would be necessary to separate granules with little pigment from highly melanized ones to compare their DNA. If melanin does indeed contain its own DNA, then it would help to explain the uniqueness of different types of melanin that are found within the deep recesses of the nervous system and the external body.

To further distinguish neuromelanin from skin melanin, Foley and Baxter (1958) found a relationship to albinism in humans. Albinos cannot produce color and primarily lack external melanin. However, a special characteristic of internal neuromelanin is suggested by the normal pigmentation of the LC and the substantia nigra of albinos who lack melanin pigments elsewhere. Likewise, European or Caucasian people have adequate pigmentation in the brain, but low amounts of pigmentation in the skin. If the amount of neuromelanin present in the pigmented neurons is species- and age-dependent (Lindquist et al., 1987), then it should be no surprise that the amount would vary among ethnic groups.

Based on the accumulated research on melanin, this author believes that high levels of melanin in and outside the body increase the likelihood of an optimal state of health. There may be further enhancements in the performance of mental activities related to movement, alertness, emotions and memory. The reader should note, however, that no notion of "superiority" is posited. What is postulated is that highly melanized individuals can reach optimal health and increase their potential more easily than those who lack melanin's enhancing capabilities.

Although we should believe that "all men are created equal," we should not blindly embrace such rhetoric. In actuality, we are clearly different in ethnicity and should not put shields over our eyes to act as though all humans are the same. From a moralistic perspective, all humans should have the same rights. However, our different physical appearances provide clear evidence that all humans are not identical. A new age of morality needs to emphasize the beauty in differences so everyone can appreciate the rainbow of humanity. Appreciating the differences would be one way to remove most bigoted attitudes that currently exist as a result of the created social construct known as "race."

In conclusion, it must be understood that the nervous system controls and integrates all bodily functions. It is the brain that controls the body and maintains stability by its countless adjustments, commands, and secretions. We have reviewed scientific research to indicate that melanin is necessary for the nervous system to function properly, and its functioning can be referred to as automatic, automatic in the sense that melanin operates as an organizing principle in the developing embryo and the nervous system, strengthening the independent viability of the developing organism.

Melanin is found in the developing embryo to assist in organizing the rapidly dividing cells. After birth, and as the infant develops, melanin is enhancing psychomotor development. During adulthood, it is the delicate balance of melanin in the nervous system that causes one to think, move, and adjust to the changing environment. Without the proper functioning of neuromelanin, the human nervous system would be deficient, and behavior would be static, i.e. lacking in automatic responsiveness or automaticity.

Next, we explore melanin's role in the endocrine system.

5

Endocrine System

Various organs in the body, especially the brain, are in constant communication with the rest of the body. The endocrine system is crucial for assisting in this intricate communication. In short, the endocrine system consists of specialized groups of tissue that can secrete substances from one bodily region to another. The organs in the endocrine system act like a network of satellites in space that transmit signals to various places on earth. The transmission inside the body takes place by chemical messengers and hormones.

A chemical messenger is classified as any substance produced by a cell of exogenous or endogenous origin that plays a physiological role in the control of cellular activity. The control can be exhibited in a proximal or distal location from the site where the chemical is produced. The following biological substances are a few of the chemical messengers that can influence cellular activity: hormones; neurohormones; neuropeptides; neurotransmitters; phermones, and growth factors.

The biological substance that is most associated with endocrinology is a hormone. A hormone can be used synonymously with a chemical messenger, and both are substances secreted by one cell to regulate another cell. Normally, the word hormone is used in the same context to describe all substances associated with classical endocrine glands (e.g., pituitary, thyroid, adrenal glands) or other cells or tissues such as the brain.

Some of the biological substances responsible for sending messages throughout the body are related to the formation of melanin. As we will see later in this chapter, some of these biological substances (e.g., melanocyte-stimulating hormone) directly influence the functioning of skin melanin and brain melanin. Since melanin is a chemical, that is how we will describe melanin's relationship to the endocrine system.

Precursor molecules are one common link between melanin and some of the biological substances secreted throughout the body. For example, the biosynthetic pathway that leads to the formation of skin melanin and brain melanin depends on a precursor molecule called tyrosine. Tyrosine has a key role in the formation of a specific category of neurotransmitters called catecholamines. In the biosynthesis of these catecholamines, tyrosine is converted to dopa to dopamine to norepinephrine and then to epinephrine. As shown in the Mason-Raper Pathway (Figure 1), however, the formation of brain dopamine (hypothesized as a preceding step to neuromelanin) and the formation of skin melanin take place in two different enzymatic pathways. The difference is found at the intermediary stage involving dopa. In the brain, dopa is converted into dopamine. In the skin, however, it is converted to dopaquinone to form skin melanin.

Unfortunately, it is believed by many scholars, specifically those not well versed in the principles of melanin formation, that the biosynthesis of catecholamines (tyrosine to dopa to dopamine to norepinephrine to epinephrine) leads directly to melanin formation. This error, made by earlier African-centered scholars (Clark (X), McGee, Nobles and Weems, 1975) has influenced many subsequent readers. Although melanin is a complex biopolymer that contains these catecholamines and numerous other chemical substances in its molecular structure, one cannot simply reason that catecholamine synthesis is responsible for neuromelanin formation. If catecholamine synthesis could be said with certainty to form neuromelanin, there should be high levels of enzyme activity in pigmented brain regions.

Since it has been experimentally demonstrated that there is minimal to no enzyme activity in the substantia nigra or any other pigmented region in human brain tissue (Rodgers and Curzon, 1975), it is inappropriate to consider neuromelanin as an end product in the biosynthesis of neurotransmitters. Moreover, if melanin in the brain is formed by some hypothetical biosynthetic pathway, the synthesis would not stem from dopamine. On the other hand, the scientific evidence states that it would have to follow a conversion similar to the Mason-Raper Pathway.

We need to understand these subtleties because they pertain to potential differences in the neuromelanin content between individuals and ethnic groups. A cautious explanation for any differences might be stated as follows: Levels of neuromelanin vary depending on an individual's overall genetic capacity to produce melanin and are not a result of the biosynthesis of neurotransmitters.

There is another class of chemical substances called peptides that are linked to melanin functioning. *Peptides* are links of amino acids. *Amino acids* are the building blocks for proteins. There are twenty essential amino acids and, depending upon how the amino acids are linked, we call the amino acid links peptides (a short chain of amino acids) or *proteins* (a long peptide chain). An *enzyme* would be the substance required to break the chain into varying lengths. To better explain the relationship between these chemical terms, the following metaphor is provided:

> Suppose you had a long chain of 250 rubberbands (protein). A small link or section (e.g., 15) of the large chain could be a peptide. Each rubberband or link would be an amino acid, and scissors (enzyme) would be needed to cleave the long chain of 250 rubberbands

49

into smaller portions.

One large protein molecule found in the brain is pro-opiomelanocortin (POMC). Several peptide messengers are found in a specific sequence of links in POMC. POMC is synthesized by cells in the arcuate nucleus of the hypothalamus and the anterior lobe of the pituitary. POMC serves as a precursor molecule from which any of several intercellular messenger substances can be cleaved.

POMC gets its name from specific chemical substances that are found in the protein molecule. For example, the opi- refers to endorphins, and endorphins (END) are the body's natural opiates or pain-killers. The melano- refers to melanocyte-stimulating hormone (MSH), primarily found in the pituitary gland. MSH is responsible for stimulating melanin production and darkening the skin. The cortin refers to the adreno-corticotropin-releasing hormone (ACTH). ACTH is one step in the biosynthetic pathway that assists the organism in physically responding to stressful conditions.

Specifically, the POMC chain incorporates many other important substances,

1) ACTH (1-39) - a hormone released by the anterior lobe that stimulates the cortex of the adrenal gland to release other hormones

2) Alpha-MSH (1-13) - a hormone released primarily from the intermediate lobes of vertebrates to darken skin. It is incorporated in the ACTH chain

3) CLIP (18-39) - a weak form of ACTH. It is found in the intermediate lobe and is also part of the ACTH chain

4) Beta-lipotropin (42-132) - a weak hormone released by the anterior lobe that causes fat cells to break down their lipids

5) Gamma-lipotropin (42-99) - a weak form of beta-lipotropin. It is incorporated in the beta-lipotropin chain

6) Beta-MSH (82-99) - Its function is undefined. It is incorporated in the gamma- and beta-lipotropin chains

7) Beta-END (102-132) - It functions as a neurotransmitter and an endogenous opiate. It is incorporated in the beta-lipotropin chain

8) Gamma-MSH - This third version of MSH lies in part of POMC and is not defined as a messenger. The ends of gamma-MSH may lie at a bond between the amino acids arginine and lysine (-57 and -56) and a bond between two arginines (-43 and -42).

Out of these elaborate sequences, we are interested in the functioning of the MSH system. Next, we will discuss the role of MSH as it is released from the pituitary.

PITUITARY GLAND

The pituitary gland is an extended brain structure that is encased in the sphenoid bone at the base of the skull. If you point a finger directly between your eyes, then you would be pointing toward the pituitary gland. Structurally, it is divided into two specialized lobes: 1. adenohypophysis; and 2. neurohypophysis. In anatomy and physiology texts, the pituitary is usually referred to as the hypophysis.

The neurohypophysis, or posterior pituitary, does not actually produce hormones, but functions in storing and releasing two hormones into the bloodstream. The two hormones are vasopressin and oxytocin. Each hormone is produced in the hypothalamus and released upon electrical stimulation from higher brain structures.

The adenohypophysis, or anterior pituitary, contains most of the hormones that regulate other organs. The numerous hormones controlling the activity of a number of endocrine glands can be found in Figure 2.

In addition to the anterior and posterior pituitary gland, there is also the intermediate lobe. The intermediate lobe, or pars intermedia, is thought to be absent in the adult human (Hadley, 1988). This absence complicates the role of MSH in humans because MSH is found in the intermediate lobe of other animals. According to Hadley, amphibians, reptiles and most mammals have MSH secreting cells in a well-defined intermediate lobe. Although the intermediate lobe is absent in certain mammals (e.g., whales and dolphins), a zona intermedia does exist in the human fetus. Despite the absence of an intermediate lobe, the zona intermedia or perhaps the anterior or posterior pituitary cells may secrete MSH (O'Donohue and Dorsa, 1982). Detection of MSH in human blood and urine and its elevation in pregnancy (Schizume, Lerner and Fitzpatrick, 1954; Clark, Thody, Shuster and Bowers, 1978) indicate that it has a role in human functioning.

CONTROL OF MSH SECRETION

The role of the intermediate lobe and MSH in the control of color changes in cold-blooded vertebrates has been documented (Hadley, 1988). It has been suggested that the hypothalamus exerts an inhibitory control over MSH release (Hadley and Hruby, 1977) to influence the color change. In this section, we will need to discuss briefly the neuroanatomical structure of the hypothalamus to understand its control over MSH secretion in the pituitary.

The hypothalamus is located directly above the pituitary gland. Since the pituitary gland is split into two lobes, there are different connecting bridges from the hypothalamus to the descending pituitary gland. To avoid becoming bewildered by the terminology, the important point to remember is that there is an intricate connection between the hypothalamus and the pituitary gland, and this connection depends on specialized brain mechanisms. What follows is an abbreviated explanation.

The hypothalamus is located at the base of the brain above the pituitary and contains more than 50 separate nuclei. The nuclei are specialized to control different brain functions and are divided into three regions in reference to the position around the third ventricle, a reservoir for cerebrospinal fluid. Immediately surrounding the third ventricle is the paraventricular zone. Juxtaposed to the paraventricular zone are the medial zone and lateral zones.

Some other terms related to the neurosecretory mechanisms are tuber cinerum, infundibular stem, and median eminence. The tuber cinerum is a location point that contains the median eminence. The median eminence is a relatively cell-free area, a saturation point for neurochemicals. The infundibular stem is part of the stalk that connects the hypothalamus and pituitary. Our concern will be to understand the neurosecretory tracts or pathways from the hypothalamus to the pituitary below.

Two neurosecretory tracts are the neurohypophysial and the tuberoinfundibular. The neurohypophysical tract is the bridge to the posterior lobe, while the tuberoinfundibular tract leads to the anterior lobe. Let us first discuss the neurohypophysial tract because it is less complex.

The neurohypophysial tract transmits vasopressin and oxytocin directly to the posterior portion of the pituitary. The special hypothalamic nuclei, in which this tract begins, are the supraoptic and the paraventricular nuclei. This tract is also called the tuberohypophysial system. Embryologically, this tract and the posterior lobe is a derivation and extension of the brain.

The tuberoinfundibular tract is responsible for the majority of hormones released into the bloodstream. Most of the neurochemicals affect the anterior lobe via the portal blood vessels. In other words, the portal system is dependent upon the tuberoinfundibular tract to release hormones into the bloodstream, and there is no other influence upon the anterior lobe. In contrast, the posterior lobe is connected by fibers to the hypothalamus. Along the tuberoinfundibular tract is the median eminence, the lowest portion of the hypothalamus. Although the median eminence is located near the posterior lobe, it projects to the anterior lobe. It is the median eminence that provides the neurochemicals needed for secretion into the bloodstream.

Dopamine is one neurotransmitter with a role in the neurosecretory mechanism. Lindley, Lookingland, and Moore (1990) have investigated two different dopamine neuronal systems that influence hormones such as MSH. Their investigation was based on experimental findings which conclude that alpha-MSH administration into the arcuate nucleus of the hypothalamus increased the release of dopamine (Lichtensteiger and Lienhart, 1977; Lichtensteiger and Monnet, 1979; Lichtensteiger and Schlumpf, 1986). Lookingland et al. stated that the arcuate nucleus contains cell bodies for the: 1. tuberoinfundibular dopamine neurons; and 2. tuberohypophysial dopamine neurons.

Tuberoinfundibular dopamine neurons terminate in the median eminence where dopamine is then released into the anterior lobe via the portal blood vessels. This dopamine is known to inhibit prolactin secretion (Ben-Jonathan,1985; Gudelsky, 1981). [Prolactin is a hormone associated with milk synthesis in lactating females]. Tuberohypophysial dopamine neurons project from the rostral arcuate nucleus of the hypothalamus to the neural and intermediate lobes of the pituitary. Dopamine released from tuberohypophysial dopamine neurons is reported to inhibit alpha-MSH and beta-END secretion from the intermediate lobe (Bover, Hadley and Hruby, 1974; Penny and Thody, 1978; Jackson and Lowry, 1983; Davis, 1986) and prolactin secretion (Ben-Jonathan and Peters, 1982).

In sum, there is evidence that inhibition of MSH secretion by the hypothalamus is controlled by a catecholamine, most likely dopamine (Hadley, 1988). In addition, catecholamines such as norepinephrine and epinephrine are inhibitory to MSH secretion from the isolated intermediate lobe (Bover, Hadley and Hruby, 1974). In the absence of inhibition, MSH exhibits spontaneous electrical activity in vitro. According to Hadley, this could be a model system for determining the relationship between cellular secretion and membrane bioelectrical activity. In other words, MSH activity is associated with the electrical activity in the nervous system that was discussed earlier in Chapter three. It was reported that melanin can exhibit bioelectronic properties, and MSH, the hormone that stimulates melanin production, can also have bioelectronic properties.

54

MSH AND MELANIN

The one unquestionable role of MSH is to control melanin pigmentation of the skin of most vertebrate species (Hadley, 1972). By considering the words in MSH—melanocyte stimulating hormones—it is clear this hormone is responsible for stimulating melanocytes to produce more melanin. Skin pigmentation will be discussed at length in Chapter eight. Therefore, the focus of this section will be confined to the relationship between MSH and melanin.

In animals, the size of the intermediate lobe is usually correlated with the animal's ability to change color. In other words, the more outstanding the ability for color change, the larger the intermediate lobe. Although we have stated that humans may not have an intermediate lobe, it has been reported that alpha-MSH may be synthesized by the pars intermedia of pregnant females (Swaab and Martin, 1981). This is an important finding because it may also suggest that the pituitary glands of ethnic groups may differ depending upon their level of skin pigmentation.

Since the pituitary gland releases hormones to regulate target organs, a similar process can be observed if we discuss the skin as an organ (see Figure 3). In fact, we can consider the skin as the largest organ of the body. After receiving input from the regulatory mechanisms of MSH, the skin releases melanin to protect the organism from the harmful effects of solar radiation. In Chapter three, we mentioned some additional functions that can elevate skin melanin beyond the primary role as a sunscreen. For example, melanin can act as a scavenger for free radicals and affect nerve impulses. Also, it is associated with the accumulation of various biogenic and metallic substances. These properties can be associated with both neuromelanin and skin melanin.

MSH guarantees that skin coloration will naturally take place when the organism must biologically adapt to surroundings. If skin coloration was not a significant phenomena, there would be no need for an endogenous regulatory mechanism such as MSH functioning.

In animals, seasonal alterations in pigment cell activity are an important aspect of the animal's ability to conform to yearly changes in environmental conditions. The control of color variation, therefore, is a dynamic rather than static process. For example, a hare is able to change color from a brown summer coat to a white winter coat. Similarly, tanning of the skin in humans takes place as a biological adaptation to the changing environment.

MSH IN HUMANS

The skin pigmenting actions of MSH are infrequently associated with the human biological system. As we have already stated, MSH is found in the intermediate lobe of most vertebrates. Since adult humans lack a well-defined intermediate lobe, there must be some other anatomical differentiation. In fact, the human pituitary contains a collection of colloid cysts and epithelial cells in the region along the border of the anterior and posterior pituitary (Herlant and Pasteels, 1967; Purves, 1966). The small number of epithelial cells which invade the posterior pituitary contain and synthesize alpha-MSH and beta-END (Celio, 1979).

Some researchers have reported very low concentrations of alpha-MSH in the human pituitary gland (Abe, Island, Liddle, Fleisher and Nicholson, 1967; Silman, Chard, Landon, Lowry, Smith and Young, 1977). Therefore, MSH was thought to be insignificant in humans. There is a significant role, however, because the pituitary of the human fetus contains a high concentration of alpha-MSH bioactivity which decreases dramatically after birth (Kastin, Gennser, Arimura, Miller and Schally, 1968; Swaab and Visser, 1977). Since the human fetus has a well-developed intermediate lobe, this may explain the high levels of alpha-MSH during pregnancy. O'Donohue and Dorsa (1982) suggest that there may be an activation of and a physiological role for the alpha-MSH system in fetal life.

Although human skin darkens within 24 hours after MSH treatment (Lerner and McGuire, 1961; 1964), Western scientists have been reluctant to declare that any extra-pigmentary functions are related to MSH in humans. An extensive review by O'Donohue and Dorsa (1982) would suggest otherwise, and a summary of the extrapigmentary effects can be found in Figure 4.

In conclusion, there is enough evidence to indicate that human MSH plays an important role in human physiology. Future research should clearly reveal that the hormonal regulation of skin pigmentation observed in smaller vertebrates is also critical in humans.

6

Visual System

The eye is the first anatomical structure involved in visual process-ing, and it is the part of the visual pathway that contains melanin. The colors we detect in our environment are dependent upon a special melinated layer of cells in the retina. In this section, we will describe how visual information is physiologically processed with the help of melanin. In this chapter, we will focus on the basic principles of vision and the role of melanin.

Every sensation the body experiences (vision, hearing, taste, feeling, smell) is activated by a receptor specialized in detection of the stimuli of the given sense. For vision, light waves are the required stimulus, and the detection of light takes place on a melinated region of the eye called the retina. The receptors in the retina are sensitive only to that tiny portion of the vast spectrum of electromagnetic radiation called visible light (see Figure 5). Although the visual pathway from the eye to the brain is quite detailed, the most important aspect of the pathway for our discussion is the retina. For clarity, a diagram of the retina and its pigmented cell layer can be found in Figure 6.

Structurally, the eyeball surveys the world from a bony socket in the skull. It is cushioned by fat and held in place by six muscles that assist in rotating the eyeball. The outer white layer is the sclera, a tough, opaque film of connective tissue. At the front, transparent tissue forms the cornea, which covers the iris. The iris has circular and radial muscles by which it expands and contracts the pupil, the dark hole in its center. When light enters the eye, it passes through the lens, the cornea, and two fluid-filled chambers: 1. the aqueous humor - small anterior chamber between lens and cornea; and 2. the vitreous humor - large posterior chamber between lens and retina. The lens, cornea and fluid-filled chambers function together as an optical system that bends light to focus and sharpen the image on the retina.

THE RETINA

The retina is a thin layer of neural tissue lining the back of the eyeball. It contains light-sensitive receptors that convert light to nerve impulses. These impulses travel to the brain by the optic nerve (cranial nerve III).

A camera and our visual apparatus function analogously. The lens of the camera and the eye function to refract the visual image. There is an iris on the camera (condenser) that functions similarly to the iris by controlling the amount of light that comes in with the image. The functional role of film in the camera is similar to the retina in the eye because both the film and the retina store the visual image.

The receptor cells (photoreceptors) in the retina are called rods and cones because of their microscopic appearance. There are about 125 million rod cells that give vision in dim light and about 7 million cone cells that give vision in bright light. A map of the retina would show rods mostly around the periphery. Near the center of the retina is where most of the cones are located. The fovea, the area of highest visual acuity, has some 4000 cones that use a separate nerve fiber. This is a unique relationship because it provides for an open and unencumbered path to the brain. On the rest of the retina, each nerve fiber serves many receptor cells, and vision is not so sharp. When you really want to focus on something of interest, the fovea goes to work.

Photoreceptors contain light-sensitive molecules called photopigments. Depending on the time of day, different forms of photopigments are activated in the rods and cones. The cones produce the sensitivity to the various wavelengths of white light that we call color vision. Rods are primarily sensitive to dim light for night vision.

Suppose you have slept outside overnight and awaken just before dawn. When darkness gives way to the sunrise, everything looks gray as the first light strikes the rod pigment, rhodopsin. The pigment bleaches as it undergoes the chemical changes necessary for seeing objects in dim light. When the morning brightens, the rods retire and the cones take over.

There are three types of cones, each receptive to wavelengths in the red, green, and blue portions of the electromagnetic spectrum. The three work in combination to create the fluctuating hues of nature. Cones, too, have light-sensitive pigments that undergo a bleaching and regenerative process.

THE PIGMENT LAYER OF THE RETINA

The pigment layer of the retina contains melanin. Melanin in the pigment layer prevents light reflection throughout the globe of the eyeball. The pigment performs the same functions in the eye as black paint inside the bellows of a camera. Acute or sharp vision is dependent upon melanin in the eye. Without it, light rays would be reflected in all directions in the eyeball and would cause diffuse lighting of the retina rather than the contrasts between dark and light spots required for formation of precise images.

Another important function of the pigment layer of the retina is its ability to influence the processing of essential chemicals required for vision. The pigment layer stores large quantities of vitamin A. Vitamin A or beta-carotene is obtained from our diet in food substances like carrots. Vitamin A is passed back and forth through the membranes of the outer segments of the rods and cones. Vitamin A is an important precursor for the photosensitive pigments, and this interchange of vitamin A is important for the adjustment of the light sensitivity of the receptors. Just as neuromelanin in the brain is found around regions that secrete substances required for normal brain functioning, melanin in the retina is associated with chemical substances important for vision. Moreover, melanin helps the physiological system store and utilize vitamins (e.g., vitamins A, C, B6 and B3) necessary for other aspects of nervous system functioning.

The major difference between neuromelanin and retinal melanin is their origin. As we stated in Chapter two, melanocytes are derived from three sources: 1) the neural crest, 2) the cranial neural tube, and 3) the outer wall of the optic cup. It is the outer wall of the optic cup that forms melanin in the retinal pigment epithelium, whereas neuromelanin is derived from the cranial neural tube.

Once again for emphasis: Melanin is found in bodily regions where energy exchange, or transduction, is occurring. In visual processing, light is converted into nerve impulses. Melanin's electrical properties assist in the transduction of light stimuli into neural messages. The strategic placement of melanin in certain locations of the body optimizes energy conversion.

In a sense, the pigment cells could be regarded as highly specialized neurons, just as the photoreceptors are highly specialized. Both melanin and photoreceptors derive from the wall of the optic cup, one from the outer and the other from the inner wall. Perhaps in considering their interrelationships we should think more of the probable "neuronal" functions and activity of the pigment cells (Breathnach, 1988).

According to Breathnach, the functions of retinal pigment epithelium are to absorb light, to prevent reflection, and to protect the photoreceptors from harmful effects. In addition to the storing and releasing of Vitamin A, melanin is involved in phagocytosis of membrane lamellae from the tips of the outer segments of the rods and cones. All of this evidence indicates that there is a functional role for melanin in vision.

ALBINO DEFECTS

The importance of optical melanin can be illustrated by its absence in albinos, who hereditarily lack melanin pigment. For example, imagine an albino entering a bright area; light would impinge on the retina and reflect off of the white surface of the unpigmented area. A single discrete spot of light that would normally excite only a few rods or cones is reflected everywhere and excites many receptors. As a result, the visual deficiencies of albinos, even with optical correction, is rarely better than 20/100 to 20/200.

Guillery and colleagues (1984; 1986) have reported interesting findings on neural abnormalities in albinos--specifically, abnormalities of the visual pathways. In mammals, they have shown that reduced melanin in the retinal pigment epithelium causes an abnormally small and uncrossed pathway. Guillery and his colleagues believe that the primary site of gene action leading to chiasmatic misrouting and an abnormal retina input into the lateral geniculate body is the retina, particularly the retinal pigment epithelium. In albinos, the line of decussation is shifted into the temporal retina, and axons from the most nasal part of the temporal retina project to the opposite rather than to the same side of the brain; the fovea is incompletely developed and projects mainly to the contralateral optic tract. As a result, there are anomalous connections in the central geniculo-cortical visual pathways. These abnormalities are linked to deficient levels of melanin in the pigment layer of the retina.

In Siamese cats, Guillery and his co-workers have provided evidence that the retinal pigment epithelium determines coordinates of the neural retina and its decussation pattern during development. A normally pigmented retinal epithelium controls the extent of the fluid space around the developing photoreceptors and the ionic content. An insufficient level of melanin during a critical stage of development can lead to physical abnormalities in the visual system. Again, then, we see the significance of melanin in the neural system--in this case, with respect to vision.

In conclusion, melanin in the pigment layer of the retina functions to enhance visual processing. One of its many functions is to facilitate the conversion of electromagnetic energy (light waves) into neural impulses. In addition, melanin can help to accumulate the necessary chemicals required to produce photopigments, without which vision would be almost impossible. Increased activity in the retina correlates into quicker reaction time and greater visual sharpness. Therefore, melanin in the pigment layer of the retina is necessary for proper vision.

7

Auditory System
and Vestibular System

The sensory mechanisms that allow us to hear (auditory) and to maintain equilibrium (vestibular) are both influenced by melanin. Although melanin is associated with both senses, not much information is reported about the specific role of this melanin. We will attempt to investigate the role in this chapter. Since the structure of the ear contains the physiological processes for both hearing and position sense, we will combine a discussion of the two senses in this chapter.

Hearing is the sense that detects sounds in the external world. Position sense or equilibrium refers to the orientation of the head in space and movement of the body through space. The ear contains two separate types of receptors for each of these processes. These specialized receptors are the site of origin for cranial nerve 8 (CN 8) or the vestibulocochlear nerve. The receptors and CN 8 serve as the sensory pathway from the ear to the brain. To describe better the hearing and equilibrium systems, we will look at the the anatomy of the vestibulo-cochlear system.

GENERAL STRUCTURE

Sound waves are the physical stimuli by which the ears detect hearing. Ears are constructed to respond to sound waves. Therefore, the hearing mechanisms must be able to transfer sound waves into nerve impulses. The head is also very important for keeping the body balanced, so equilibrium senses are strategically located in the ear, which has three parts: 1) the external ear, 2) the middle ear, and 3) the inner ear. A diagram of the ear can be found in Figure 7.

The External Ear

The external portion of the ear consists of the pinna and the external auditory meatus. The pinna is composed of cartilage and a covering of skin. To some extent, it serves to collect sound waves and direct them into the external auditory meatus. This function is of little value in humans, but it is of greater importance in some lower animals. The external auditory meatus or auditory canal extends inward from the pinna to the tympanic membrane (ear drum), a distance of approximately one and one-half inches. It acts like a resonating column to channel the larger sound waves into smaller and smaller waves.

The Middle Ear

The middle ear transmits sound waves from the external ear to the inner ear. It is a tiny, irregular cavity in the temporal bone. It contains three auditory ossicles or bones: 1) the malleus, 2) the incus, and 3) the stapes. The names of these very small bones describe their shapes as a hammer, anvil and stirrup, respectively. The bones vibrate together to exert control over sound waves. Since the inner ear is fliud-filled, the force of the sound wave must be amplified in the middle ear in order to overcome the fluid inertia of the inner ear.

The middle ear is filled with air, and there are several openings into the middle ear cavity. One opening is covered by the tympanic membrane leading from the auditory canal. Another opening is the eustachian tube, which is a passage way to the upper part of the throat. The other two openings consist of the oval window and the round window, which lead to the inner ear. The clinical importance of these middle ear openings is that they provide routes for infection to travel. In addition, the eustachian tube serves to equalize pressure in the middle ear when the atmospheric pressure changes (e.g., ears "popping" when the altitude changes).

The Inner Ear

The inner ear is the part of the vestibulocochlear apparatus on which melanin has its greatest effect. It is also the most complex part of the ear. We will briefly describe the structure of the inner ear and then explain how melanin can affect hearing and position sense later in this chapter.

The inner ear is also called the labyrinth because of its complicated shape. It consists of two main parts, a bony labyrinth and, inside this, a membranous labyrinth. The bony labyrinth consists of three parts: 1) the vestibule, 2) the cochlea, and 3) the semicircular canals. The membranous labyrinth consists of the utricle and saccule inside the vestibule, the cochlear duct inside the cochlea, and the membranous semicircular canals inside the bony ones. The membranous labyrinth can be divided into the auditory portion and the vestibular portion. The auditory portion is called the cochlea and the semicircular canals and vestibule make up the vestibular portion.

The fluid which fills these membranous structures is called endolymph. The general structure of the inner ear might roughly be compared to an automobile tire and its tube. The tough outer casing of the tire is comparable to the bony labyrinth and the rubber inner tube to the membranous labyrinth.

1) The Cochlea

The word cochlea, which means snail, describes the outer appearance of this part of the bony labyrinth. When histologically sectioned, the cochlea resembles a tube wound spirally around a cone-shaped core of bone, the modiolus. The modiolus houses the spiral ganglion, which consists of cell bodies of the first sensory neurons in the auditory relay to the brain.

Located within the spiral tube are several important structures that produce the sensation for hearing. The membranous cochlear duct resembles a lopsided triangle as it lies within the bony cochlea. It extends like a shelf across the bony canal to the sides of which it is attached. The portion of the bony canal above the cochlear duct is called the scala (meaning "stairway") vestibuli, and the portion below is called the scala tympani.

The roof of the cochlear duct is thin and is called Reissner's membrane or the vestibular membrane. The floor is composed of the basilar membrane. Resting on this basilar membrane is the spiral organ of Corti, the essential receptor and organ for hearing. It is a very complicated structure, consisting of a supporting framework on which rest the important hair cells. These fine, delicate structures are called hair cells because each one of them bears about 20 fine hairs on the basilar membrane. Above the hair cells is a fixed tectorial membrane. Vibration of the spiral organ of Corti results in the hairs being bent against the above tectorial membrane. As a result, nerve impulses initiated by the stimulated hair cells pass over dendrites whose cell bodies are located in the spiral ganglion in the modiolus. Axons extending from this ganglion unite to form the cochlear branch of CN 8.

2) The Vestibule

The vestibule is a disk-like cave in which float two membranous sacs, the utricle and the saccule, each of which is filled with endolymph. On the floor of the utricle is a patch of hair cells which constitute the vestibular receptors or maculae. Entangled in the brushlike hairs are tiny particles of calcium carbonate that are called otoliths. Nerve impulses are initiated when the otoliths push or pull on the hairs and bend them in the direction of the pull of gravity. The vestibular branch of CN 8 transmits impulses to the cerebral cortex, and we recognize the position of the head in space as it relates to the pull of gravity.

3) The Semicircular Canals

There are three semicircular canals in each ear. They are arranged at right angles to one another so that all three planes of space are represented. Lying within these bony tunnels are the membranous semicircular ducts which loop through semicircles, beginning and ending at the utricle. At one end, each duct has a dilatation called the ampulla, which contains a patch of hair cells or receptor end organs known as the cristae. Movement of the endolymph bends the hairs and sets up impulses in the vestibular branch of CN 8. The cristae are stimulated by sudden movements, or by a change in rate or direction of movement. In other words, whenever movement begins and ends, accelerates or decelerates, or changes in direction, nerve impulses are set up in dendrites leading to CN 8.

HEARING

Hearing requires that sound waves are transduced by the auditory system. Sound waves must be projected through air (outer ear), bone (middle ear), and fluid (inner ear) to stimulate nerve endings and set up impulse conduction over CN 8. Nerve impulses are then sent to the auditory area of the brain for further processing.

Sound waves in the air enter the auditory canal and strike the tympanic membrane (ear drum) of the inner end of the canal. Vibration of the ear drum moves the middle ear bones, and a cascading effect causes the malleus, the incus and the stapes to vibrate. The last bone in the relay, the stapes, is attached against the oval window and the vibration moves the fluid in the cochlea. When the stapes moves against the oval window, pressure is exerted inward. This starts a rippling effect in the fluid that is transmitted through Reissner's membrane (the roof of the cochlear duct) to endolymph inside the cochlear duct. The ripple is then transmitted to the basilar membrane (the floor of the cochlear duct) and then to the organ of Corti, which is supported by the basilar membrane.

From the basilar membrane, the ripple is next transmitted to and through the perilymph in the scala tympani and finally expends itself against the round window. At this point, hair cells of the organ of Corti transmit nerve impulses to nerve cells connected to the spiral ganglion, which leads into CN 8. The movement of the hair cells against the adherent tectorial membrane stimulates neural impulses along CN 8 to the brainstem. Before CN 8 reaches the auditory cortex in the temporal lobe, nerve impulses must pass through "relay stations" in nuclei in the medulla, pons, midbrain and thalamus for further processing.

EQUILIBRIUM

The maintenance of equilibrium depends on the position of the head and special mechanisms within the vestibule of the inner ear. For example, located within the utricle and the saccule lies a small structure called the macula. It consists mainly of hair cells and a gelatinous membrane that contains small particles of calcium carbonate called otoliths. A few delicate hair cells are embedded in the gelatinous membrane, and receptors for the vestibular branch of CN 8 contact the hair cells of the macula located in the utricle. Changing the position of the head causes a change in the amount of pressure on the gelatinous membrane and causes the otoliths to pull on the hair cells. This stimulates the adjacent receptors of CN 8, and the subsequent nerve impulses to the brain to produce a sense of the position of the head and also a sensation of a change in the pull of gravity (acceleration and deceleration).

If a person is rotated in a chair, the endolymph in the semicircular canals is set into motion and stimulates hair cells. A sense of movement either of self or surroundings (vertigo) occurs, together with a peculiar movement of the eyes called nystagmus. Nystagmus consists of a rapid movement of the eyes in one direction and a slow movement in the opposite direction. These sensations create a situation known as motion sickness to make one nauseous.

In sum, it may be stated that the semicircular canals are dynamic sense organs. They rouse sensations of starting and stopping movement, changing its speed and direction. They also originate many reflexes which involve skeletal, smooth and cardiac muscles. The utricle is a static sense organ. It gives information regarding the orientation of the head in space, in relation to gravity, and sets into action postural and righting reflexes. Injury to the semicircular canals or utricle leads to disturbances in maintaining one's orientation and balance. Vertigo and nystagmus are common signs of vestibular pathology.

INNER EAR MELANIN AND
THE VESTIBULOCOCHLEAR SYSTEM

Inner ear melanin has been scientifically investigated since the mid 19th century. Pigment cells, resembling those of the choroid coat of the eye, were discovered in 1851 by Corti in the labyrinth of the cow and sheep. Later investigations demonstrated the presence of pigment in the cochlear spindle of the human ear (Voltolini, 1860). As a result of Voltolino's findings, Waldeyer (1871) and Ranvier (1875) both found similarity between the pigment cells in the eye and the ear. Since 1931, the pigment in the inner ear has been known as inner ear melanin (Wolf, 1931). Some relationships that have been noted over the years are the amount of melanin in the inner ear is related to the pigment amount in the iris (Bonaccorsi, 1965), and significant differences in the temporary threshold shift have been found between brown- and blue-eyed individuals (Tota and Bocci, 1967).

Inner ear melanin is embryologically derived from neural crest cells, and it is widely distributed in the cochlea and in various parts of the membranous labyrinth (Hilding, 1977; LaFerriere, 1974). In the cochlea, melanin is present in the modiolus and spiral lamina. In the membranous labyrinth, melanocytes are found in the utricle, saccule, ampulla, and in the endolymphatic duct and sac (Breathnach, 1988).

The utricle and the ampulla are each connected to semicircular canals. The melanocytes of the utricle and ampulla are located subepithelially below the basement membrane between connective tissue cells (Meyer zum Gottesberge, 1988).

The inner ear is another one of those hidden regions of the body in which melanin migrates to during the darkness of embryological development. What possible role could melanin provide for such a secluded sensory system like the inner ear? We have already discussed that the middle ear bones function to amplify sound waves in the middle ear. As the sound waves travel toward the inner ear, melanin can also assist the amplification process. What it may do is enhance the conduction of nerve impulses. It is proposed that inner ear melanin is performing its role as an extrasensory perceptive molecule. In humans, there is an abundance of inner ear melanin in the modiolus and the endolymphatic sac. Remember, the modiolus houses the first sensory neurons for CN 8. Therefore, inner ear melanin is located right at the conjunction site for the initiation of nerve impulses for hearing. It may amplify the subsequent neurons in the auditory relay to the brain. Moreover, melanin may be found in the endolymphatic sac to assist in the maintenance of ions in the endolymph fluid. This would be a very important role for nerve conduction.

In the vestibular organ, inner ear melanin may help synchronize hearing and the rhythmic movement of the body. This synchronization is very important for staying in tune with the melody of a song and rhythmically dancing to the tune. From the ancient African drumbeats to the modern African rap songs, inner ear melanin has probably played a key role in producing sensational styles of music such as rhythm and blues, jazz, rap, hip-hop etc. The majority of people of European ancestry lack the natural rhythm possessed by most people of African descent. If skin pigmentation has any correlation with the amount of inner ear melanin, then differences in natural rhythm may be explained accordingly.

Singing a song, snapping your fingers, bopping your head and wiggling your hips all at the same time can be a rudimentary task for some people and a bewilderingly complex task for other people. The rhythmical talent possessed by African-American entertainers such as

Michael Jackson and M.C. Hammer shows the limitless potential that people of African descent can reach. To this author's knowledge, there are no Caucasion or Asian entertainers that can compare to the raw talent possessed by African-American entertainers like Jackson and Hammer. It is more than just "hard work" that could produce a Michael Jackson or a M.C. Hammer. It is the belief of this author that raw talent is a product of an optimally functioning physiological system.

In addition to musical entertainment, African people can also excel beyond other ethnic groups in the areas of athletics. We do not intend to suggest all forms of athletics, but certainly the case is strong when we refer to sports such as basketball and football. These physically demanding events require advanced functioning from internal melanic systems. Michael Jordan may be used as an example when we explore this chapter's focus on the inner ear mechanisms that attribute to advanced physiology. I emphasize that the discussion of Michael Jordan and other players is made with specific reference to inner ear melanin.

Although hearing is important, the vestibular mechanisms make inner ear melanin a major factor for producing highly versatile athletes. During rapid movements, reflex activity will adjust the position of the head with respect to gravity and the horizontal plane. As a result, the body is adjusted relative to the position of the head to affect normal postural alignment. These reflexes originate partly from the vestibular system. As I mentioned earlier, the vestibular system consists of: 1) semicircular canals which are stimulated by angular acceleration of the head, and 2) the utricle and saccule which are stimulated by linear acceleration and gravity.

The vestibular system has direct neural connections to the eyes, spinal cord and cerebellum. These connections allow the individual to regulate changes in head position. The vestibular system, however, is not the only system responsible for providing information about body orientation. Vision, touch, proprioceptive information from muscles and joints, and neuromelanin in the brain's motor system also play key roles for body orientation. We are only emphasizing the role of inner ear melanin in this section.

As players like Michael Jordan go up to score a basket, they may need to change their body position instantaneously to avoid a blocked shot. In milliseconds, the ball may need to go behind the back, under one leg and dunked with one hand to beat the opposition. A move such as this requires advanced physiological mechanisms. Melanin probably plays a key role by enhancing the firing of nerve impulses in the vestibular system. It is vitally important that the player land safely and on balance.

People of African descent have revolutionized basketball because of their extrasensory perceptive play. Before people of African descent were allowed to officially play in professional basketball leagues, the games were played on the horizontal plane with jump shots and one handed lay-ups. Currently, there is no description to state how blacks have moved beyond the horizontal plane to another dimension. Likewise, football players of African descent have revo-lutionized football. The awesome speed and balance of running backs like Gale Sayers and O.J. Simpson to the acrobatic receivers like Lynn Swann and Jerry Rice depend upon a highly efficient physiological system. In order to accelerate in the air from full speed, leap higher than the defender, catch the ball with one hand, and land on one foot and keep running one must have well-developed vestibular mecha-nisms. Although there are excellent athletes from European or Asian descent, the most dynamic and versatile athletes are those highly melinated people of African descent.

There are some other possible roles of inner ear melanin that are worthy of mentioning. Consideration of the general biophysical and biochemical properties of melanin has further led to speculation that inner ear melanin may serve as a reservoir for trace elements. In other words, it may act as a sink for free radicals (Breathnach, 1988). It may also provide cytoprotection against ototoxic drugs and high-intensity noise damage.

Meyer zum Gottesberge (1988) has looked at the physiology and pathophysiology of inner ear melanin. In her brief review, special attention was drawn to the composition of melanin and its presumed function as a biological reservoir for divalent ions and as an ion exchanger. In addition, it may play a role as an intracellular buffer for calcium ions.

In terms of pathophysiology, Meyer zum Gottesberge stated that it is not clear why some people develop severe hearing impairment while others retain normal auditory threshold when exposed to the same level and equal duration of noise. Perhaps decreased cochlear pigmentation is related to the severity of noise-induced hearing loss. It has been experimentally shown that individuals with a brown iris have shorter temporary threshold shifts than individuals with a blue iris or those who are albinos. This result supports McGinness et al. (1974) who theorized about melanin's ability to be a threshold switcher. In addition, investigations on industrial related noise-induced hearing loss suggested that individuals with dark eyes suffer less than those with lighter colored eyes (Meyer zum Gottesberge, 1988).

It is known that hypopigmentitis is associated with defected ocular systems as well as auditory systems (Creel, O'Donnell and Witkop, 1978; Creel, Garber, King and Witkop, 1980). Meniere's disease is an example of how hypopigmentation is associated with hearing loss. The leading symptoms of this disorder are tinnitus, vertigo, and fluctuating hearing loss; however, the etiology of Meniere's disease is unknown. Meniere's disease is unknown among black South Africans and it is very rare in Uganda and West Africa (Nsamba, 1972). However, it is common in the white population. Inner ear melanin may be related to the differences in musical creativity and musical preferences between ethnic groups. If there is clearly a physiological difference, then inner ear melanin may be an important determinant in the response to acoustic stimulation (Meyer zum Gottesberge, 1988). Melanin may have a protective function whereby sound energy peaks are cut off through excitation of melanin and later converted to thermal energy (McGinness et al., 1974: Lyttkens, Larsson, Stahle and Englesson, 1979).

The nature of black music is very rhythmic. It is like the steady beat of the heart. The thump of the heart and the thump of the bass in black music (gospel, rhythm and blues, hip-hop, jazz, rap, etc.) are harmonically driven. The acoustic sound of the bass may convert to thermal energy and cause highly melinated individuals to become saturated with feelings. These feelings can be transformed into the rhythmic movements that are created by all types of black music. Inner ear melanin may cause a highly excited state of energy transformation. In contrast, people who are deficient in melanin may prefer high-pitched and erratic music (rock-n-roll, heavy metal, etc.) in order to feel the vibrations. The fact that black people can dance rhythmically to the thumping bass of black music while others may exhibit uncoordinated movement when dancing to rock-n-roll or heavy metal is evidence to me that inner ear melanin is an extrasensory perceptive molecule possessed mostly by black people and critical for acoustic stimulation.

To conclude, we cannot overemphasize the complexity of the vestibulocochlear system. The location where melanin is primarily found is the inner ear. At this site, melanin can simultaneously influence both hearing and equilibrium. In effect, an optimally functioning melinated inner ear can cause a person to be efficiently rhythmic and well coordinated. If we can surmise that people of African descent are genetically programmed for all types of melanin to function at a greater capacity, then we can scientifically validate why people of African descent are more rhythmically coordinated when compared to other ethnic groups.

8

The Skin and Its Hormones

The importance of the skin should not be underestimated. It is the outside covering of every internal process and mechanism of the body, the barrier between our internal functions and the outside world. If we view the skin as one large organ, we can say it secretes a hormone (melanin) that gives color to the exterior portion of the body. We will explore the uniqueness and versatility of the skin in relation to melanin functioning in this chapter.

First, we will discuss the structure of the skin to describe its overall importance to the body. Secondly, we will review many functional roles provided by the skin. Thirdly, we will demonstrate how skin pigmentation can be influenced by exogenous substances called psoralens. Since the skin is actually a functioning organ, we will discuss the skin's ability to produce, manufacture, and regulate hormones. My aim is to show that the skin has many functional roles besides its established role as an outer covering for protection from the external environment.

STRUCTURE OF THE SKIN

The skin is a tough, pliable covering for the body consisting of two layers: 1) the epidermis--a surface layer; and 2) the dermis--an underlying thicker layer. The epidermis is a tough covering about one millimeter thick. It consists of a mass of dead cells that are constantly worn away and replaced by cells emerging from deeper layers. These deeper cells push outward, but as they move farther away from the nourishing capillaries below, they begin to die. The protein material of the dying cells changes form, becoming a new material called keratin. Ashy, flaky, dry skin is observed when the epidermal cells are sloughed off. The epidermis has several cellular layers such as the stratum corneum, the stratum lucidum, the stratum granulosum and the stratum germinativum.

The dermis constitutes the greater part of the total skin thickness. The range of thickness is from 0.5 mm over the eyelids to 6 mm over the upper back, neck, palm of the hand and sole of the foot. The dermis is divided into a superficial papillary layer, which contains many capillaries. The capillaries bring blood into proximity with the germinating layers of the epidermis. A deeper layer, the reticular layer, is a dense mass of interlacing white and yellow elastic connective tissue fibers. The reticular layer accounts for the toughness and strength of the skin. Since the dermis contains fat, numerous blood vessels, nerves, sensory receptors, hair follicles, and sweat and sebaceous glands, it is known as the true skin.

FUNCTIONS OF THE SKIN

A. *Protective Role*

The skin can provide protection from the environment in two ways. First, the sebaceous and sweat glands provide a thin film over the entire skin surface. The sweat glands secrete water, lactic acid, amino acids, urea, uric acid and ammonia. The sebaceous glands contribute fats, fatty acids, and wax alcohols. These secretions enable the film to act as an antiseptic and a neutralizer of acid and alkai, to interfere with absorption of toxic materials, to act as a lubricant of the horny layer, and to control hydration of this layer.

Secondly, the keratin from the horny layer is an almost complete physical barrier to electromagnetic waves, bacteria, fungi, parasites, and practically all noxious chemicals. Keratin is a tough, fibrous protein that has a tightly packed arrangement of cells to prevent the movement of molecules through the intact epidermis.

B. *Thermoregulation*

The sweat and sebaceous glands can act as a large radiator surface to handle the continuing adjustment in heat loss and heat production. Since the blood is the primary medium of transfer of heat throughout the body, the complex vascular patterns found in the skin play a significant role in the regulation of heat loss. In addition, thermal sensory receptors located in the skin, stimulated by heat or cold, are connected with the central regulating mechanisms in the brain.

C. *Excretory Function*

Accumulated substances need some way of being released from the body. The skin contains sweat and sebaceous glands, which allow substances to be released from the body. The sweat and sebaceous glands excrete different types of substances, however.

Sweat glands can be divided into two types called eccrine and apocrine. Eccrine glands, or small sweat glands, are distributed all over the surface of the body, excluding the lips, glans penis, and clitoris. The eccrine glands are innervated by the CNS and respond strongly to sensory and psychic stimuli. For example, the eccrine glands in the skin of the palms and soles respond slowly and weakly to heat, but more strongly to stimuli that induce nervousness and fear.

Apocrine glands, or large sweat glands, do not occur over the whole body, but are found in the armpits, around the nipple of the breast, on the abdomen around the belly button, in the perineal and perianal regions, in the prepuce and scrotum of the male, in the perigenital region of the female, in the external ear canal and in the nasal passages. Apocrine glands function periodically and respond promptly to mental stimuli. In general, they do not respond to thermal stimulation, and their secretions do not take part in thermoregulation.

Sebaceous glands are different from the sweat glands due to their different secretions. Sebaceous glands produce an oily substance called sebum, which forms the greater part of the lipid component of the surface film covering the skin surface. Sebum anoints the hair and keeps it from drying and becoming brittle. Sebum forms a protective film on the surface of the skin that limits the absorption and evaporation of water from the surface. In humans, sebaceous glands protect the skin surface where hair follicles are present. They are relatively small on most areas of the trunk and extremities, but quite large in the skin of the forehead, face, neck and upper chest.

D. *Absorptive Function*

The waxy surface film on the skin and the tightly packed cells of the horny layer provide efficient barriers against the penetration of most substances through the skin. However, experiments have demonstrated that the skin is permeable to substances that are lipid and in the gaseous state. Depending on the substances, the absorptive function can help beautify the skin or cause debilitating problems like cancer.

The lipid-soluble barrier of cells is constructed to ensure that sufficient amounts of water permeate the cells. Since the membrane is made of a lipid bilayer, lipid-soluble substances are able to pass through cell membranes rather easily. Phenols, hydroquinones and salicylic acid (aspirin) are lipid-soluble compounds that enter the skin with extreme ease from any vehicle. Phenols would be compounds in which a hydroxyl group is bonded directly to a specific carbon atom of an aromatic ring. A typical phenol is more acidic than an alcohol, but less so than a carboxylic acid. When the hydroxyl group of phenols are oxidized, they form hydroquinones. Hydroquinones such as dopaquinone are very critical in the formation of melanin.

79

Other lipid- or fat-soluble substances are estrogen, testosterone, progesterone, and deoxycorticosterone. Lipid-soluble vitamins that penetrate the skin with ease are A, D, and K. Of the lipid-soluble salts of heavy metals, mercuric bichloride is absorbed most easily through the skin. Salts of lead, tin, copper, arsenic, bismuth, antinomy and mercury, which are originally lipid-soluble, can be absorbed through the skin. For example, they can be transformed on or in the cornified layer into lipid-soluble substances by forming oleates in combination with fatty acids of the sebum.

With the exception of carbon monoxide, substances that are in the gaseous form at ordinary temperature easily penetrate the skin. The amount and rate of transfer depends on three factors: 1) the vapor tension of the gas inside and outside of the skin; 2) the temperature; and 3) the water and fat solubility of the gas. Gases transmitted easily through the skin are oxygen, nitrogen, helium, carbon dioxide, ammonia vapor, hydrogen sulfide, hydrogen cyanide, and volatile aromatic oils. Physiologically, the movement of oxygen and carbon dioxide is most important.

E. *Skin Pigmentation*

Pigmentation of the skin is the most pertinent function related to the science of melanin. In previous chapters, we have covered some of the essentials of melanin functioning by discussing the role of melanocytes, melanocyte-stimulating hormone, and some physiological effects associated with melanin. In this section, we will look at melanin's most common role as a pigment.

Melanin is commonly known as the pigment that functions to color the skin and, more importantly, to protect humans from the hamrful effects of ultraviolet radiation (UVR). Melanin is contained in specialized structures called melanosomes. Melanosomes are synthesized by cutaneous cells of neural crest origin called melanocytes. Melanosomes synthesized by melanocytes are transferred to keratinocytes, inside of which they are transported to the epidermal surface. Ethnic differences have been demonstrated in the size, numbers, and arrangement of melanosomes within the multilayered epidermis. Also, it is the multicellular epidermal melanin unit (melanocyte and associated pool of keratinocytes) rather than the melanocyte alone that is the focal point for melanin metabolism in the human epidermis (Quevedo, Fitzpatrick, Pathak and Jimbow, 1975; Witkop, Quevedo and Fitzpatrick, 1983).

Melanocytes are multidendritic cells located at the epidermal-dermal junction of the skin. Although the population densities of epidermal melanocytes vary from one bodily region to another, all humans, regardless of skin color, have approximately the same numbers of melanocytes. Ethnic variations in skin color, however, result from differences in the properties of melanosomes and not from differences in numbers of melanocytes.

The principal pigments of vertebrates are the melanins. According to Pawelek and Korner (1982), there are five basic colors in mammalian skin and hair: black, brown, red, yellow and white (the absence of pigment). Different combinations and intensities of these colors are found in a variety of hues and patterns that can be passed from generation to generation. Other researchers state that the major sources of color are melanins (black-brown) and carotenes (yellow) of the epidermis, and oxyhemoglobin (red) and deoxyhemoglobin (blue) of the dermal blood vessels (Quevedo, Fitzpatrick and Jimbow, 1985). The black-brown melanin of the epidermis and hair is designated as eumelanin, and the yellow-red melanin as pheomelanin.

Solar radiation (UVR, visible light, infrared light) and UVR from artificial light sources stimulate the production of melanin. First, melanin is incorporated into a subcellular structure referred to as a premelanosome. When this organelle is fully melanized, it is known as a melanin granule or melanosome. The epidermal unit, mentioned earlier, contains the melanosomes that pass through the multidendritic processes of the cells from which they are released into the surrounding cells of the epidermis.

The biosynthesis of melanin is highly dependent on tyrosine and related enzymes. Tyrosine is an essential amino acid utilized by melanocytes to synthesize melanin. The first observations to implicate tyrosine in the biosynthesis of melanin were discovered in mushrooms (Bourquelot and Bertrand, 1895). The authors observed that the cut flesh of mushrooms, when exposed to the air, turns red and then black. A similar effect can be observed when a cut apple, potato, or banana turns brown.

After Bourquelot and Bertrand boiled the mushrooms in ethanol, a colorless substrate was extracted which reacted with the oxidizing ferment of the mushrooms to yield first a red and then a black compound. The oxidizing ferment was the enzyme tyrosinase, and the red and black compounds were dopachrome and melanin, respectively. The compound extracted from mushrooms by ethanol was soon identified as tyrosine.

The classic version of skin melanin biosynthesis can be demonstrated in the Mason-Raper Pathway (see Figure 1). In sum, tyrosine is oxidized into dihydroxyphenylalanine (dopa) by tyrosinase. Dopa is oxidized by the same enzyme to dopaquinone, which rearranges spontaneously to leucodopachrome and then to dopachrome. A subsequent intermediate of dopachrome is 5,6-dihydroxyindole-2-carbolic acid. It loses its carboxyl group to become 5,6-dihydroxyindole. Upon exposure to air, 5,6-dihydroxyindole is oxidized to indole-5,6-quinone, and then into melanochrome, a purple compound which polymerizes to melanin.

The Mason-Raper Pathway is specific for skin melanin formation. It is not as clear, however, whether or not brain melanin forms similarly. It is commonly believed that neurotransmitter synthesis leads directly to melanin formation. According to the Mason-Raper Pathway, melanin is not an immediate by-product of neurotransmitter synthesis. As a result, a different explanation for the formation of brain melanin is needed.

In order to make sense of our explanation on the biosynthesis of melanin, it should be understood that melanin is a large and ubiquitous molecule. The melanin molecule is composed of certain neurotransmitters and other biological substances. Therefore, one can assume that the structures embedded in the melanin molecule are directly responsible for melanin formation. Some possible pathways involving neurotransmitter synthesis of brain melanin are found in Figure 8.

As I said in the introduction and in Chapter five, however, brain melanin is not solely dependent on the enzymic processes that are responsible for the synthesis of neurotransmitters. Although brain melanin and brain catecholamines coexist in humans, brain melanin, specifically, is found in certain brain regions due to genetic influences on nervous system development. In other words, brain melanin is genetically programmed for distribution to certain brain regions independent of a specific process involving enzymes.

EIGHT-METHOXY PSORALEN

Darkening a person's skin without the direct effects of sunlight is not a new idea. The scientific investigation of such an effect can be linked to ancient Africa. For example, ancient Africans were aware of a natural substance that could be ingested to darken their skin (Ronan, 1973; Harber and Bickers, 1989). It is reported that ancient Africans chewed a root called ami-majos when they went on caravans across the desert sands. This root provided extra protection from the sun by reinforcing skin pigmentation. What is fascinating about this phenomenon is that modern research shows that the root contains the active organic chemical compound called 8-methoxypsoralen (Ronan, 1973; Fitzpatrick, Hopkins, Blickenstaff and Swift, 1955).

A concise description of 8-methoxypsoralen (8-MOP) can be found in the *Physician's Desk Reference*. In addition to melanizing agents, there are demelanizing agents that can be found in this reference. Moreover, there are three available journals of current 8-MOP research: 1) *Journal of Investigative Dermatology*; 2) *Photodermatology*; and 3) *Pigment Cell Research*.

8-MOP is a potent photosensitizer of the skin. It is used in combination with ultraviolet radiation (320 nm to 400 nm) to increase skin tolerance to sunlight, to facilitate repigmentation in vitiligo, and to treat skin diseases such as psoriasis, eczema and mycosis fungoides (Epstein, 1990). After oral ingestion of 8-MOP, the patient is exposed to a specially designed ultraviolet lamp system. Increased sensitivity of the skin to such radiation appears in one hour, reaches a maximum in two hours, and disappears in about eight hours. Exposure of 8-MOP-treated patients to ultraviolet light thickens the stratum corneum, induces an inflammatory reaction in the skin and increases the amount of melanin in the exposed area.

Although it is common medical practice to administer ultraviolet light with 8-MOP treatment, 8-MOP can stimulate the melanocyte system without ultraviolet radiation (Walther, Haustein, Rytter and Gast, 1989; Marwan, Jiang, Castrucci and Hadley, 1990). In other words, one could take a pill and become "black" in a matter of hours without the direct effects associated with sunlight.   B l a c k n e s s (e.g., viewpoints, ideas, culture, etc.) or dark skin color (i.e., people of African descent) is not looked upon favorably in this race-conscious world. Therefore, it is highly unlikely that 8-MOP is being intensely studied by white scientists with the goal of making white people black or darker. Since people of African descent have no problem producing melanin, the findings from 8-MOP research can have profound consequences for people who lack melanin. Besides psoriasis and other skin disorders, a more important reason for this investigation may be for protection against sunlight.

Presently, the earth is in a state of crisis and there are major threats to the earth's environment. For example, the release of chlorofluorocarbons into the atmosphere is believed to be thinning the ozone layer that protects living things from dangerous ultraviolet rays. The ozone layer, located in the stratosphere, between 10 and 30 miles up, is vital to the well-being of plants and animals. Ozone molecules, which consist of three oxygen atoms, absorb most of the ultraviolet radiation that comes from sunlight. Over Antarctica, the ozone has at times been depleted by as much as 50 percent.

The dangers of ozone depletion will inevitably affect white people more adversely. Besides excessive amounts of utlraviolet radiation that could reach the earth, the small amount that does get through to the earth's surface inflicts ample damage. In animals and humans, the rays have been linked to sunburn, cataracts, and weakened immune systems. Ultraviolet light carries enough energy to damage DNA and thus disrupt the operations of cells. Thus, excessive exposure to sunlight is thought to be the primary cause of some skin cancers (Lemonick, 1989).

Man has always shown a great capacity for adjusting to change. Past generations have survived floods and ice ages, famines and world wars. But when dealing with the environment, there is grave danger in relying on adaptation alone; societies could end up waiting too long (Elmer-Dewitt, 1989). People of European descent have had a proclivity to try to control nature, and the scientific manipulation of skin pigmentation is another example. Through experimentation, darkening properties of 8-MOP could be processed and mass-produced for a global emergency. In particular, it could be used to protect white-skinned people from the harmful rays of the sun if the ozone continues to diminish. Although the space program is making progress towards the potential existence of human life on another planet, mass distribution of 8-MOP may need to occur before there can be an alternative colony in outerspace.

Other uses for melanizing agents can be seen in the sport of body building. White body builders darken their skin to enhance their physical definition. The dark colored skin increases their chances for winning body building competitions. Tanning salons are useful, but the effects of ingestable melanizing agents may be less dangerous. In addition to skin pigmentation, there is an attempt to link 8-MOP to behavioral functioning.

The link between psoralens and behavioral functioning is the hormone melatonin. Melatonin is a nocturnal hormone that is released from the pineal gland. In animals, it has been associated with the cycles of reproduction and the changing of skin color. In humans, it is known to retract melanin granules and is associated with mood disorders such as seasonal affective disorder. It has been demonstrated that psoralens increase melatonin levels without ultraviolet light (Grota, Lewy, Goldsmith and Brown, 1985).

Grota and others (1985) studied the effects of 8-MOP on melatonin levels in rats. They found that serum melatonin levels were elevated in both intact and pinealectomized (extracted pineals) animals relative to vehicle controls (received no 8-MOP). The greatest increase was in the intact and not the pinealectomized animal.

In the following years since the study by Grota and others, there has been an attempt to study the behavioral influences in humans. A good example is the study conducted by a French team of researchers. In a series of experiments led by Eric Souetre, the effects of psoralens have been linked to dementia, depression, abnormalities associated with circadian rhythms (Souetre, Salvati, Krebs, Belugou and Darcourt, 1989; Souetre, Salvati, Belugou, Robert, Brunet and Darcourt, 1988) and the general increase of melatonin levels in humans (Souetre, Salvati, Belugou, 1987; Souetre, De Galeani, Gastaud, Salvati and Darcourt 1989; Souetre, Salvati, Belugou, Kreb and Darcourt, 1990).

The French team has extensively studied the effects of 5-MOP (another form of psoralen) on normal subjects and mentally ill patients. In sum, chronic (one-week) and one time administration caused an elevation in plasma melatonin. In a one-week trial with 5-MOP administered daily (40 mg/daily) at 0900 hr. to normal subjects, stimulation of melatonin secretion was enhanced and more pronounced when the drug was administered during the dark phase of the day (Souetre et al., 1990).

In another study (Souetre et al., 1989), they investigated the melatonin response to 5-MOP in healthy adult and elderly subjects; and in demented, depressed, and schizophrenic patients. The subjects' plasma melatonin levels were evaluated before they took single, oral 40 mg doses of 5-MOP, and again, three hours later. The findings demonstrated that the melatonin response of the pineal gland to 5-MOP is reduced in demented patients compared to that of healthy elderly subjects. Depressed and schizophrenic patients had a normal melatonin response to the drug. The authors stated that the pineal gland may be functionally impaired in dementia but not in depression.

Future research should help clarify the role, if any, that psoralens can have in behavior. Presently, the clear effect of psoralens is the ability of these compounds to darken skin color. Individuals suffering from specific skin disorders and white- skinned people who are directly harmed by sunlight can find the photochemotherapeutic effects of psoralens very beneficial.

SOLTRIOL (VITAMIN D)

In this section, we will explore an innovative conceptualization of our pigmentary system and vitamin D synthesis. New terminology will be applied. For example, we will refer to vitamin D as a steroid hormone because to call it a vitamin is a misnomer. We will use the name soltriol (Stumpf, 1988) to describe this sun-derived hormone. In addition, we will look beyond the restrictive roles that have been placed on some biological functions. In terms of restrictive roles, soltriol is thought to affect only calcium absorption, and melatonin is believed to affect only reproduction in mammals. In fact, soltriol and melatonin work in a complimentary fashion to correlate biological activities with the daily and seasonal changes of our solar environment.

A. Skin Color Transformation

The first step in explaining skin color transformation is to analyze how pigmentary transformation can take place. Extensive literature has already been researched by Charles Finch (1990; 1991). I will attempt to summarize Finch's research in order to shed light on the discussion of melanin, melatonin, and soltriol synthesis.

The first humans originated on the African continent, and they were unquestionably dark in skin color (black). Reinforcing the fact that the original humans were black should help the reader understand the pigmentary transformation that has occurred over time. Speaking in terms of evolution, black parents gave rise to brown, white, red, and yellow pigmented humans all over the world.

Early human fossils were found around the equatorial region in Africa. Any organism born near the equator has a dark exterior. The organism will be lighter in color the further it is from the equator. Finch (1991) informs us that this phenomena is known as Gloger's Law. Gloger's Law states that the closer an animal is to the equator, the darker the coat; the closer to the Arctic Circle, the whiter the coat.

The dark exterior is a result of the pigment melanin. Melanin provides protection against the harmful effects of ultraviolet radiation. However, melanin can be disadvantageous in regions of the earth that have limited sunshine. The development of rickets (bone deforming abnormality) is a prime example of how melanin can be maladaptive.

Rickets develops when there is insufficient conversion of soltriol in the skin. Soltriol, in turn, facilitates the absorption of calcium to make the bones strong. Depending on the latitude, therefore, the skin must be the right complexion to properly regulate soltriol synthesis.

As early dark-skinned humans migrated out of Africa, they settled in many geographical regions. Natural climatic changes such as the ice age caused migrating humans to be separated from other early humans for long periods. Over time, humans had to adapt to their environment in order to survive. Dark melinated skin was a natural adaptation in Africa, but became a hinderance the further they went from the equator. During the ice age, environmental conditions were horrendous; often no sun shone at all, and when it did it was limited. Therefore, early humans had to adapt physiologically to cold geographical conditions.

Evidence that dark skin was a problem in places where the sun barely shone can be found in human fossils with rickets. Limited sunshine and dark skin equal less vitamin D, which is primarily responsible for helping the body absorb calcium for strong bones and teeth. Therefore, the body had to adapt to the changing environment. At this point, we can say that humans genetically transformed from black to white. An important point to remember is that the human body has internal mechanisms that naturally attune the body to daily events, like sunrise and moonrise, in our solar system. These internal mechanisms ensure the survival of humans under evolving planetary conditions.

In addition to this ancient history, Holick (1985) reported related evidence dating back to the eighteenth and nineteenth centuries. During the industrial revolution in Northern Europe, people began emigrating from rural areas to congested industrial centers. Children played in dank alleyways surrounded by the pall of industrial pollution. As early as the mid-seventeenth century, European scientists recognized that young children living in such places were afflicted with a malady that caused severe muscle weakness and rickets. These phenomena closely resemble the processes that occurred thousands of years earlier.

Our physical bodies are in constant communication with the surrounding universe. It is only natural that the human body has physiological functions that respond to and depend on the sun and other cosmic phenomena. If we understand that man is a microcosm of the macrocosm, then it is simple to see how our bodies are intimately connected to nature. The rising and setting of the sun and the cyclical relationship between the sun and moon are critical for all living systems.

B. *Soltriol and the Sun*

The sun is the generator of all life processes on earth. We can never underemphasize the importance of the sun or neglect the power of its radiance. If the sun did not shine, humanity and all living organisms would cease to exist. We have already demonstrated, through the example of the photosynthetic pathway, the sun's critical role in maintaining life. By observing the relationship between nature and man, we see that the sun can directly influence two pigmentary systems: 1) the pigment chlorophyl in plants; and 2) the pigment melanin in humans. This relationship between the sun and organic life illustrates the importance of the sun to the microcosm (the human body) and to the macrocosm (the earth, or universe).

The sun has played a significant symbolic and religious role in the culture of many civilizations. Centuries ago, the indigenous peoples of the North American continent built mini pyramids, called Etowah Mounds, in praise to the sun. These Mounds still stand and can be observed in northern Georgia. Ancient African people in Kemet, or what is now Egypt, praised the sun's power with the name RA. RA was not worshipped as a god but was a connection between human-kind and the sun. In a sense, the sun is like a drug in that it can produce hormones -- vitamin D, for example. Symbolic representations of the sun and its relationship to man still exist on ancient temples and papyri. Some figures depict the sun's rays touching the forehead and skin of a human being. Current research provides evidence that the pineal gland, located in the center of the brain at the level of the forehead, is the organ regulated by the sun to secrete the hormone melatonin. This explains even more vividly the intimate cycles of nature, skin tone, and melanin functioning.

C. *The Endocrinology of Sunlight and Darkness*

In Chapter five, we saw that the endocrine system consists of specialized groups of tissue that secrete substances (hormones) from place to place in the body. In this section, we are interested in the endocrine hormones (melatonin and soltriol) that regulate melanin. In addition to regulatory mechanisms by which hormones control melanin, we will consider potential behavioral effects associated with the endocrinology of sunlight and darkness.

Soltriol is widely believed to be "the calcium homeostatic steroid hormone" that acts peripherally with calcitonin and parathyroid hormone to regulate calcium levels in blood and tissues. The effect on calcium, however, is only one role of soltriol. Another perspective is that soltriol can act as a comprehensive somatotrophic activator and modulator. This new concept proposes that the steroid hormone soltriol is the skin-derived endocrine messenger of sunlight, functioning to coordinate the life processes of an organism's internal milieu with conditions in our solar environment. Soltriol may be a significant hormone helping to optimize an organism's development, growth, and reproduction (Stumpf, 1988).

Furthermore, melatonin is one of the major pineal hormones that have been linked to mammalian reproduction. Suppression of gonadal activities in relation to the daily and seasonal dark-light cycles is generally believed to be the main action of melatonin. In humans, melatonin functioning has been linked to mood disorders such as Seasonal Affective Disorder (SAD). Since there are binding sites for melatonin in several places in the central nervous system, there is a much more complex role than the regulation of reproduction.

Biochemicals in the body do not sit around and wait to function like a bullet in a gun waiting to be fired. These numerous chemicals undergo regular daily rhythmic fluctuations in accord with our solar environment. Antagonistic or complimentary effects for both soltriol and melatonin function during different times of the day. In relation to the sun, soltriol and melatonin are the respective endocrine hormones for sunlight and darkness.

The effect of sunlight on soltriol formation is regulated in the skin by melanin content and distribution (Stumpf and Privette, 1989). The role of melanin in the regulation of soltriol production can be described based on observations that the "hormone of sunlight," soltriol, stimulates melanin production in vitro (Tomita, Fukushima and Tagami, 1986), and that the "hormone of darkness," melatonin, lightens melanocytes (Lerner, Case Takahashi, Lee and Mori, 1958). It is suggested that this dual regulation produces an interactive relationship between the two endocrine systems in which soltriol is complimentary to the effects of melatonin.

From autoradiographic studies, it has been demonstrated that there are wide distribution sites for soltriol and melatonin binding in the nervous system and endocrine glands. There are strong potential influences of soltriol and melatonin on neural activities in diverse components of the central nervous system with effects on motor, sensory, endocrine and autonomic systems. The presence of melanin in these key areas and the skin can significantly influence the functioning of these complimentary hormones.

As an intermediary substance, melanin uses the sun to generate soltriol and possibly other steroid hormones. Stumpf and Privette (1989) stated that soltriol, like other steroid hormones, has target sites in the brain and therefore can be expected to have organizational effects on the developing brain. Furthermore, soltriol can induce permanent changes as well as activational effects on the mature brain, and cause transitory functional alterations. Organizational effects may include entrainment of certain biorhythms, while activational effects may include heightened mood, increased alertness, extrasensory perception and other responses and functions yet to be determined.

In sum, sunlight is important for our bodily functions. Eating a balanced diet with natural foods, drinking plenty of pure water, exercising and getting exposure to sunshine (depending upon your hue) can enhance your health. Melanin is a critical link between our human bodies and the elements of nature, enabling us to optimize our health.

9

Melanin and Early
Childhood Development

In Part I, we reviewed developmental embryology and the location of melanin at the earliest stages of human development (i.e., gestation). If melanin were not important for cellular and behavioral organization, then internal regions of the body, where there is no sunlight, would not be black. Clearly melanin does more than protect against ultraviolet radiation. To strengthen the argument for melanin's advanced functioning, we presented scientific evidence (see Chapter three) demonstrating that melanin has many functions in our physiology. In this chapter, we will establish connections between early childhood development and advanced melanin functioning.

Many studies on fetal development have been conducted by William Smotherman and Scott Robinson at the Center for Developmental Psychology at SUNY Binghamton, NY. Their findings revealed that fetal movement in the intrauterine environment is not uncoordinated or only reflexive according to outside stimulation, but is behaviorally organized (Smotherman and Robinson, 1990). According to these authors, the behavior of mammals has its root in the prenatal period. A complete understanding of behavioral development must include investigation of the behavioral capabilities of the fetus (Smotherman & Robinson, 1988). With advanced techniques, it has been observed that the fetus is more than a passive passenger during gestation; it is an active organism responsive to changes within its intrauterine environment.

Behaviorally organized and coordinated movements in adults depend on melanin and dopamine in the substantia nigra located in the midbrain. It is obvious that adult humans need organized physical activity in order to survive, so it is not surprising that the fetus acts to adapt to its intrauterine environment. Smotherman and Robinson (1990) state that fetal behavior is inextricably connected to the environment in which the fetus has developed, to the environment in which it exists at a given moment, and to the succession of predictable environments that will follow in the course of its life history. The organizing molecule (Barr, 1983) responsible for these prenatal origins of behavioral organization is melanin.

Although fetal development is highly dependent on important factors such as the mother's genetic structure, nutritional intake, and mental and physical status, the blackness of melanin plays a significant role in forming a viable human organism. Since we know that some people produce more melanin than others, differences in melanin production may explain why African (black) infants have enhanced physiological and psychomotor development.

African and European people are at opposite ends of the spectrum of humanity's diverse population. Clear differences in childhood development are observed. For example, infants of African descent are born with advanced sensorimotor skills compared to European infants.

Amos Wilson, a psychologist and prolific writer, has written extensively on the black child. Two of his books, dealing with research and theoretical viewpoints, are intended to help black parents cultivate the advanced development of their children: *Developmental Psychology of the Black Child* (1978); and *Awakening the Natural*

Genius of Black Children (1991).

In these books, Wilson reviews numerous studies that compare psychomotor and physiological development of black and white children. Research has consistently demonstrated that up to at least the first two years of life, black children are significantly more advanced than their white counterparts (Smart and Smart, 1972). Black children have a "natural head start." According to Wilson (1991), if black children are appropriately stimulated, motivated, guided and supported, they are capable of attaining intellectual heights equal or superior to the children of any other ethnic group.

An earlier study by Geber in 1958 produced revealing results. Geber demonstrated that African infants were superior to European infants on several indices of psychomotor development. Within a span of a few hours after birth to 11 months, Geber shows that psychomotor development of African infants was twice as fast as that of European infants. The psychomotor development of European infants probably lagged because of a less melinated, and therefore less efficient, nervous system. Some of the behavioral indices were:

1. Being drawn up in a sitting position without head falling backwards;

2. Supporting oneself in a sitting position and watching reflection in the mirror;

3. Standing against a mirror;

4. Climbing the steps.

Pierce (1980) stated that "this sensorimotor learning and development were phenomenal...a superior intellectual development held for the first four years of life."

The native African children studied were from Uganda and Kenya. Ugandan children between six and seven months old demonstrated object identification in a toy-box retrieval test in which the experimenter showed the infant a toy in a tall box. The child then walked across the room to retrieve the toy. European and American children were not expected to do this until somewhere between 15 and 18 months (Geber, 1958). Geber concluded that this precocity was in both motor development and intellectual development.

In Kenya, Leiderman (1973) studied infants of the Kikuyu people. The Bayley scales of infant development were used to assess physical maturation and perceptual and sensory functions. Kikuyu children scored higher than the standard score for American infants of the same age. A comparison of Kikuyu infants with black and white infants in America and the United Kingdom revealed that Kikuyu children performed better on both mental and motor tests throughout the 15 month period of observation.

If we take an empirical look at childhood development from an international perspective, we see a clear difference between black and white children. For example, Jackson and Jackson (1978) stated that the greatest mental and physical precocity has been found in black infants, both in Africa and in the United States, followed by Indian infants in Latin America and infants in Asia. European infants rate lowest on the precocity scale.

Furthermore, Bayley (1965) compared the psychomotor development of 1409 infants between one month and 15 months in 12 cities across widely scattered areas of the United States. It was demonstrated that black children showed higher values for every month at every level except 15 months when compared with their white counterparts. In another study, Walter (1967) compared black and white infants across socioeconomic lines. The results demonstrated that black infants were generally superior to white infants in psychomotor development. A very plausible explanation for these findings is that melanin is advancing the development of black (African) infants (Stewart, 1981).

We want to emphasize that melanin functioning is the key determinant for enhancing early childhood development. Since low- income teenage mothers who eat malnutritious meals can still produce a child with advanced sensorimotor skills (Wilson, 1978), we can eliminate socioeconomic and dietary factors as key determinants for early psychomotor development. An explanation, however, must be given for the decline in these early developments.

Wilson (1991) has stated the following reasons why the "natural head start" of many black children is not continued beyond early childhood:

1. The cumulative effects of nonstimulating interaction.

2. The cumulative effects of extremely young maternal age on children's intellectual development.

3. The cumulative effects of the poor quality of the home environment.

4. The cumulative effects of inadequate language and vocabulary-mastering experiences.

5. The cumulative effects of inappropriate educational and schooling experiences.

In conclusion, we have provided evidence that melanin in the human organism is critical for enhancing psychomotor development during infancy. However, melanin cannot overcome all societal conditions that can thwart, stagnate, negate or reverse development. With the appropriate stimulation, motivation, support and guidance, parents of black children can eliminate negative factors that "under-actualize" the remarkable intellectual potential of black children (Wilson, 1991).

Throughout Part Two, we have researched the scientific literature in order to detail the science of melanin, and in order to show that melanin has crucial physiological roles beyond protecting humans from ultraviolet radiation. In Part Three, I will explore the metaphysical connection related to melanin functioning and offer ideas for future research.

Part Three - Future Outlook

Nonwhite people are generally believed to be more in tune with nature by living in peace and harmony with the world and its inhabitants (leading to spiritualism). White people, however, have generally been more inclined to control and dominate nature (leading to materialism). Cosmologically, white and nonwhite people view the world differently, and it is primarily due to cultural influences and environmental conditions. In other words, ethnic differences are the result of different patterns of human progression over thousands of years of being separated in specific geographical regions of the earth. Some people have evolved from cold and resource-limited environments, whereas others have evolved from tropical and resource-plentiful environments (Diop, 1974). Depending on the environment, one can develop thought patterns centered around materialism or spiritualism. In the following chapter, we will relate the material and spiritual realms of existence to our topic on melanin. Also, we will explore new horizons for the future of melanin research in Chapter 11 and conclude this book with suggestions for future scientific research into the science of melanin.

10

Material-Spiritual Connection

The nervous system and skin are two essential biological systems linking the internal milieu of humans to the exterior world. They act as specialized vehicles to allow humans to maintain contact with the material and spiritual realms. It is proposed that melanin is the molecule, both in the human body and in the universe, that acts as a conduit between the visible (material) and the unseen (spiritual) world.

The material plane consists of tangible substances, while the spiritual plane consists of metaphysical experiences beyond physical reach. Although substances in the material plane can be physically manipulated, the spiritual plane can only be perceived. There is an intricate connection between the material-spiritual planes, and the melanin in the skin and nervous system helps to provide metaphysical experiences.

The Universe

The universe is the totality of existence. Just as our bodies evolved from a single cell called an embryo, the universe is a system created out of one entity. More clearly, the human body is a microcosm of the macrocosm, a replica of the surrounding universe. Here is a list of natural phenomena and biological elements that symbolize each phenomenon:

1) Spiral Milky Way - Spiral DNA molecule

2) 9 planets - 9 holes in the body - 9 major endocrine glands

3) MELANIN in the sky - MELANIN in the skin

4) High content of H_2O compared to mass on earth and in the body

5) Divine universal consciousness interconnected with mental consciousness

The following scenario will assist the reader in understanding metaphysical principles that are often elusive. First, imagine yourself standing in an open field at 12 o'clock midnight. The sky is dark but illuminated by the stars, moon, and other celestial bodies. Your visibility is not impaired by clouds; nothing hinders direct communication with the universe. Your body is positioned to transmit and receive the energy around you, though this energy is invisible to the physical eye.

Modern science demonstrates that there are electromagnetic forces, light and sound waves emitted in the universe. Therefore, there is, quite literally, energy around us that we cannot see.

Another form of energy that is not well understood pertains to consciousness. For example, consciousness itself emits energy that dwells in the universe. Psychologically, a person's conscious or consciousness refers to a state of awareness of what is happening within or around oneself. A person who lacks consciousness would be thinking on a bestial level.

Here is a metaphor to better explain the transmission of conscious waves. A radio has been designed to transmit sound waves into what we hear on the radio. What we hear depends upon the channel, or the signal accessed. The different channels correspond to frequencies or amplitudes of sound waves. Consciousness works in a similar fashion.

A person's level of awareness is associated with the programmed content of a person's mind. Therefore, evil-minded people will be attracted to situations associated with evil. Good-natured people will be attracted to positive situations. Certainly, a person can oscillate between the extremes of good and evil, but physical and mental status are what determine a person's ability to "change channels" and tune into different levels of consciousness. According to Alder (1968), each thought is a vibration with a measurable wavelength.

The meditation process is a way a person trains the mind and body to tune into divine universal consciousness. It is interesting to note that many cultures have theories on creation. Usually, a simple story or philosophical concept, explains the origin of the universe. The common denominator tends to be, quite logically, that the world as we know it originated from a thought generated by the creator. In fact, every thought we have originates in a being whose life is given by a first creator, and is therefore, an extension or continuation of the creation of the universe itself. Thus, we "conceive" our world; we create our own individual worlds, lives, and ourselves.

The ultimate goal of meditation is to reach divine consciousness, referred to as "nirvana" in Buddhism. Once people attain "nirvana," or some other stage of exceptional spiritual consciousness, they are no longer fascinated and trapped by material things, and are free to transcend the material world and relate to everything in a spirit of oneness. This unity with the divine consciousness in our universe can be explained by physical science.

The atom is the basic unit of all molecules in our universe. Any material item (i.e., tree, book, car, shoe, jello, helium, ice, etc.) that exists in this world vibrates. When you break down all the different forms of substance in the universe, they vibrate on the same level. Even though the hardness, softness, gaseousness, liquidity, etc., of substances are different, their atomic structures vibrate on the same level. Moreover, the Dogon people of West Africa believe that the entire universe is moving; man on earth is in motion and life, even inside the smallest seed of grain, is in movement as well (Griaule and Dieterlen, 1986).

Vera S. Alder (1968) investigated ancient wisdom and established these findings regarding vibration:

> "Both the ancient sages and modern scientist are agreed that everything in life is formed of vibrations...We are told that vibration is the result of force or energy, concentrated in some mysterious way and caused to vibrate, shake or oscillate at different speeds. The composition of an atom...is a tiny vacuum, round which this force or energy revolves as a vortex, just as the circle of the sun's aura or zodiac revolves around it. The zodiac contains the planets revolving within it, and the minute 'zodiac' of the atom contains also its planets, or electrons as they are called...The difference between one object and another is ultimately a question of vibration." p. 26.

The relationship between vibration and the wholistic nature of the universe is currently being investigated in physics, The Grand Unified Theory (GUT).

The GUT is a way for Western scientists to explain the metaphysical principles familiar to people in less technologically advanced cultures. Western scientists have reached a point where they are no longer dealing with purely physical or chemical things, but have pushed upwards through the great scale of vibrations into the clouds of conjecture (Alder, 1968). The GUT hypothesis states that during extreme high-temperature conditions, the universe's early moments, quarks, and leptons were interconvertible. Quarks are particles with combinations of electrical charges, and leptons are particles that interact with each other only through electromagnetic and weak forces (Golob and Brus, 1990).

As the temperature of the universe cooled and energies were less available, the quarks and leptons began to acquire distinct identities. According to the GUT, however, the transformation of quarks into leptons—and vice versa—remains a theoretical possibility. In sum, modern physicists cannot explain how the present variety of forces and subatomic particles could have evolved from an initially homogeneous universe.

The concept of consubstantiation presented by Nobles (1986) is another way of looking at the fact that everything is made up of an original substance. Therefore, meditation, or the awareness that everything in the universe is connected, can give one a feeling of unity with the creator.

John A. West (1987) has written a very informative book to help us deal with these metaphysical principles. He shows a connection between vibratory phenomenon and health. According to West, an analysis of vibratory phenomenon and the physical body can be observed in the stress and strain of everyday life. For instance, our complex environments take a toll on our emotional and physical faculties. According to West:

> "People go haywire living too near an airport or working in the incessant noise of a factory. Office buildings that recirculate air and make extensive use of synthetic materials create an atmosphere depleted of negative ions. Though undetectable by the senses directly, this is ultimately A VIBRATORY PHENOMENON ON THE MOLECULAR LEVEL,* and it has powerful, measurably harmful effects: people become depressed and irritable, tire easily and lose resistance to infection. The subsonic and ultrasonic frequencies produced by a wide variety of machinery also exert a powerful and dangerous influence." p. 37.
> * Emphasis added.

West continues:

> "...the daily life of city dwellers today is technically a form of mild but persistent torture, in which victims and victimizers are equally affected...The result is similar to that wrought by deliberate torture. The spiritually strong recognize the challenge, meet it and surmount it. The rest succumb, become brutalized, apathetic, easily swayed...and men are easily moved to violence or to condoning violence in what they imagine to be their interests. All of this brought about by men professing high ideals, but ignorant of the forces they manipulate." p. 38.

Understanding the nature of the universe and our intimate connection to the cosmos is paramount in terms of influencing human and other natural life forms. Depending on the nature of the forces in the universe, our bodies can either utilize energy in a positive way or be adversely affected. Next, I will further explore the theory of universal connection.

A Universal Connection

Our nervous system and melinated skin are specifically designed to harness energies circulating in the universe, and the brain is a transmitting and receiving device for this invisible energy. Although radios and televisions have been developed to pick up pre-existing waves of energy, the brain has evolved to detect the pre-existing conscious waves. West (1987) postulates:

> "The brain is not an `evolved' organ which, somehow, accidentally over the aeons generated a faculty we call 'consciousness'. Rather, it was developed in order to receive and apprehend those aspects of universal consciousness necessary for man to perform his foreordained task - just as the radio does not generate radio

waves but receives the waves already there. The nature of consciousness determines the structure of the brain; the nature of sound determines the structure of the ear; the nature of light determines the structure of the eye..." p. 137.

With regard to the intimate links between microcosm and macrocosm, West states the following:

"Man is the sum of the principles that pervade and organize the universe; he is the self-perfecting product of the grand experiment that is organic life on earth, embodying within himself the mineral, vegetable and animal kingdoms. His body is the temple designed to permit him to carry out the rite of self-perfection - the only legitimate human goal. All other goals lead to apathy or disaster, as is obvious from any daily newspaper." p. 135.

He continues:

"Man is a model of the universe. If he understands himself perfectly, he also understands the universe: astronomy, astrology, geography, geodesy, measure, rhythm, proportion, mathematics, magic, medicine, anatomy, art...all are linked in a grand dynamic scheme. No aspect can be isolated from another and treated as a separate specialty or field without distortion and destruction." p. 136.

Most of West's ideas came from ancient Kemet (Egypt). He has studied ancient African thought and believes, like other Western authors (Schwaller de Lubicz, 1977; Schwaller de Lubicz, 1979), that this early thought was superior to the belief systems of today. His research provides evidence that scientists in ancient Kemet created a body of knowledge to help humans master and harness invisible forces in nature. Indirectly, modern science in the area of electromagnetism is providing scientific evidence that there is much to learn about the many forces of energy in the universe.

Electromagnetism

The fact that we have an electromagnetic spectrum informs us that there are invisible forces of energy in our environment. The word electromagnetic refers to the nature of the radiation emitted from electromagnetism; it is both electrical and magnetic at the same time and capable of immediately becoming either. For instance, the field around a magnet can attract or repel properties. A similar field is found around anything that is electrically charged. The static electricity from a rug or the varying oscillating waves that make up a radio transmission are examples of electromagnetic fields in electricity.

Electricity and magnetism are interconvertible. In other words, energy can move back and forth from each state. Magnetism, for example, can be turned into electricity by simply moving a conduction wire in its field. The television broadcast is one example of what happens when you spin a coil of wire between the electromagnets of a generator. What is intriguing about receiving signals for a television broadcast (or radio) is the construction of the antenna.

The antenna is specially constructed to receive emitting waves of energy. For example, an erect wire of correct length on your roof can detect electromagnetic waves and pick out a pulse related to its own length. The waves are then turned into a varying electric current. The converted electromagnetic waves reach the electronic equipment and translate into television pictures (Redgrove, 1987). An antenna is specially constructed to receive radiant energy from the man-made world of electronics. Of more importance, however, is the natural world where all electromagnetic waves begin. The body responds to man-made as well as natural electromagnetic waves.

African people naturally have kinky or wiry hair. Non-African people have matted or animal-like hair. In other words, kinky or wire-like hair is an evolutionary advance since very few animals (sheep,

106

buffalo, yak etc.) have hair similar to African people. Kinky or wire-like hair is constructed like an antenna to absorb more readily those naturally occurring electromagnetic waves in nature. The "conscious waves" we mentioned earlier are also more attracted to kinky or wire-like hair. Matted or animal-like hair may generate electricity in the form of static; however, it is matted and limits the conversion of radiant energy into other forms of energy.

In nature, electromagnetic waves are being turned into electric currents in conductors like copper wire or iron ore deep in the earth. These currents flow and create magnetic fields, which in turn charge moving conductors with a current to send out an electromagnetic wave. This conduction relates to our prior discussions on melanin. Melanin is formed by the conversion of tyrosine by tyrosine hydroxy-lase (tyrosinase). Purified tyrosinase contains copper and has been shown to be a mixture of three isoenzymes (Pawelek and Korner, 1982). Copper in the earth and the body function similarly to assist in the transference of energy. Moreover, melanin is found in both the earth and body.

The electromagnetic spectrum consists of radiant energy classi-fied according to specific wavelengths that range from very short cosmic rays present in interstellar space and longer wavelengths utilized in radio and television communication. The sun is the natural source for most of the electromagnetic radiation in our atmosphere. A mixture of oxygen, water, and carbon dioxide molecules form the ozone layer, which filters out a significant portion of solar radiation. In Chapter eight, we discussed the hazardous effects of ozone deple-tion. The biological threats of the electromagnetic spectrum can be counteracted by the absence of this filter or by an over abundance of electromagnetic radiation inside the ozone. Many forms of radiation may have beneficial therapeutic effects. If used improperly or because of accidental exposure and pollution, electromagnetic radiation may prove hazardous.

Another polemical issue about electromagnetism is its source. For example, the Western world's most notorious astronomers believe that the universe is electromagnetically sterile. James Granger (1988) has researched literature by Velikovsky (1977) and Schell (1982) to state otherwise. Essentially, Granger has ridiculed Western science and the perspective of a steady state Newtonian concept of the universe, which holds that gravity determines the movements and orientations of heavenly bodies.

In his book called *MO'*, Granger lambasts astronomers like Carl Sagan. Sagan (1980) stated that solar systems are held together by gravitational forces and atoms by electrical forces. Granger questions this statement by asking, "If atomic particles - protons, neutrons and electrons - have mass, and gravity is a force that is intrinsic to mass, then shouldn't this mysterious force called 'gravity' have an effect on atoms? Why does it apply to the celestial arena and not the atomic arena?"

With regard to melanin, an electromagnetic arena may explain how melanin is functioning in the universe. For example, space is an electromagnetic web containing rotating masses of solids, liquids, and gases; and electromagnetic forces play an important role, influencing the movement of these elements. These elements form planets and stars, and these interstellar masses spin or rotate to generate an electromagnetic field around themselves. If we look at electromagnetic radiation as a force, we see how stars and planets are giant generators.

Furthermore, the positive and negative charges emitted from electromagnetism produce attraction and repulsion. Forces of attraction and repulsion provide the potential for disturbing the electromagnetic web in space to produce disorder among stars and planets. The concept of space as an electromagnetic web means that cosmic collisions and/or cosmic creation are the result of disorder in the universe. Rather than viewing space as void or empty, Granger speculates on a plasma (dust and gas) in space that is capable of carrying electromagnetic charges. It is proposed in this book, that the plasma in space referred to by Granger is dark matter, or melanin.

Western scientists have the technical equipment needed to investigate melanin in the universe, but are bewildered by the findings. They realize there is something out there but cannot quite figure it out. Trefil (1993) remarked:

"Dark matter is strange stuff. It's all around you but you can't see it. It's whistling by your ears but you can't hear it. It is arguably the most important material in the universe, but until recently, scientists had no idea that it existed. It will decide the fate of the universe, but we have no idea what it is. How can scientists after so many centuries, still know so little about the working of the cosmos?"

Although white/Western (Eurocentric) thinkers are dumbfounded by the mysteries of outerspace, most nonwestern cultures have concepts and activities to express their understanding of the cosmos. In fact, nonwestern scientists would probably not be amazed by the sudden "discovery" that interstellar space is not empty or void.

In sum, the special bioelectronic properties of melanin are critical for the energy that is constantly transmitted to and from outerspace. It is this dark matter (melanin) that maximizes the connection of melinated human bodies to the spirit of the universe. Through the elaborate world of electromagnetism, our skin and nervous system link our internal melanin to the cosmic melanin to allow us to tune in with cosmic nature.

Chaos Theory

In this chapter, I have discussed the use of the Grand Unification Theory (GUT). It is a theoretical viewpoint created by Eurocentric thinkers to explain the holistic nature of the universe. Chaos Theory is another Eurocentric formulation devised to explain another phenomenon (randomness) that is difficult to interpret when one holds linear Newtonian views of the world. Both the GUT and Chaos Theory are theoretical attempts by Western scientists to explain phenomena that are beyond the scope of logical deduction and reasonable thinking. In this section, I will introduce the reader to Chaos Theory and speculate on the role of melanin in our random and fluctuating world.

In a series of articles in the journal, *Science*, Robert Poole (1989a;

1989b; 1989c) has written about Chaos Theory and how it relates to disorder and randomness in certain physical and biological systems. It is a mathematical concept rather difficult to define precisely, but can be thought of as deterministic randomness; "deterministic" because it arises from intrinsic causes, not from some extraneous noise or interference. Randomness refers to irregular, unpredictable behavior (Poole, 1989c).

Historically, Eurocentric thinkers believed the world was orderly and clock-like. Everything could be explained by observing, manipulating, and controlling the environment. For these reasons, it is easy to see why modern Eurocentric thinkers have primarily been concerned with forces that are tangible or materialistic. Their way of viewing the world has only been able to explain things that can be reasoned; things that are materialistic and nonspiritual. If the phenomena cannot be controlled or manipulated and explained rationally, they give it little credence. Furthermore, the phenomena must be able to be counted and measured, otherwise arguments are put forth against its very existence.

Inability to predict phenomena is a serious problem for Western scientists. Rational-minded thinkers thrive on controlling phenomena, and prediction is an essential element to reasoning and thinking about a problem. Losing the ability to predict and control phenomena has sent shock waves throughout Western science. The formulation of Chaos Theory is Western science's way of saying, "We cannot explain everything."

Presently, Chaos Theory is used to study complex phenomena in nearly every scientific discipline. Chaos allows order to be found in such diverse systems as the atmosphere, dripping faucets, or the heartbeat. Chaos Theory is being used to explain things that standard science cannot, and it offers new tools to study irregular behavior (Poole, 1989c). It is a type of randomness that appears in certain physical and biological systems and is intrinsic rather than caused by outside noise or interference. Chaotic systems operate under a wide range of conditions and are, therefore, adaptable and flexible. This plasticity allows systems to cope with an unpredictable and changing environment.

The organizing capabilities of melanin (Barr, 1983) may directly influence certain physical and biological systems. Since Chaos Theory emphasizes that order is found in disorder, we can speculate that the ubiquitous presence of melanin and the physical properties of this substance provide a sense of order in a seemingly chaotic world.

From a nonwestern point of view, Dogon philosophy states that knowledge lies in knowing man (Griaule and Dieterlen, 1986). Moreover, all that is not man is obtainable knowledge because man has been given the ability to know that which is not himself. The Dogon people believe that the world is conceived as a whole. This whole was thought, realized, and organized by one creator, God, in a complete system which includes disorder. The Dogon seem to have understood the meaning of chaos before modern Eurocentric thinkers realized that the universe is not clock-like.

Next, I will review an intricate system in the human body that is the driving force behind metaphysical experiences. I will discuss the nature of the Chakra system and how it functions as a mediator between the chaotic mesh of the spiritual and material realms of existence.

The Chakras

Earlier in this chapter, I referred to several types of energy forces in the universe. It is possible for this energy to be received by the human body in certain locations of the head and torso (see Figures 9 and 10). The locations or vital force centers are known as Chakras. Similarly, there are locations on earth that are known to be centers of attraction for cosmic energy (e.g., pyramids in Egypt, Stonehenge in England). If one can understand how the essential polarities of spirit and matter come into existence through vital force centers, then it will be much easier to analyze the relationship between the universe's energy and the chakras.

Chakras, as well as auras and electromagnetic fields, are as old as the earth itself (Bruyere, 1991). The chakra system can be considered as an energy system which keeps body and mind alive and healthy. The word chakra is most commonly used to describe the energy system. It is a sanskrit word meaning "wheel of light." The wheels of light emanating from the chakra system are manifested between the opposing poles of spirit and matter.

In man, the essential polarity of opposites has its axis along the spinal column in which spirit has its manifestation at the crown of the head while matter (in its densest form) manifests at the base of the spine (Rendell, 1979). Spirit and body are the basic polarities, and thought is the relationship between the two. In relation to the Dogon belief that the conception of the whole world was a first thought, only a disciplined person can attain divine universal consciousness. The intertwining of mental consciousness (thought) and divine universal consciousness is maximized during meditation. The different levels of consciousness we referred to earlier pertain to the life force vibrating at different frequencies, and chakras respond to different levels of vibration.

In any system, a potential difference between the two poles gives rise to a flow between them. The nervous system contains the framework for conducting the flow of electricity in the body. In electrical terms, this potential difference is a voltage, and the amount of flow between the poles can be measured as current. The current flows up the spine spirally like DNA and the Milkyway.

The ions we obtain from digestible substances give our bodies the power to generate electricity. As our nervous systems generate electricity, there is a flow of current that creates a magnetic field around the body. In man, the polarity of spirit and body gives rise to a magnetic field which surrounds the body in the form of an aura.

Auras are often spoken about in general conversation, however, there are additional metaphysical terms that describe physical fields around a person's body. In order to explain how energy in the universe could affect human behavior via the chakra system, we will briefly explain seven planes or states of matter that are known to make up our physical bodies (see Figure 11).

Everything in life, from celestial bodies to pebbles on earth, are interpenetrated by seven planes of existence (Alder, 1968). For example, man possesses a body made up of solids, liquids, and gases. These three physical elements make up three material states in the

Physical World. Outside of the physical body is a fine web through which the electric and radiating life forces are fed into the physical body from the outer universe. This fine web, in which the chakras communicate, is known as the *Ethereal World.*

The next plane is the *Astral World.* Emotions and desires are manifested in this sphere, stirring and motivating us. It is a world of attraction and repulsion. The world of thought or mind supersedes the astral world in the form of the *Mental World.* All levels of mental consciousness fluctuate throughout the mental world to give us the ability to gain knowledge. It is important to note that thought is not confined to the brain; the brain merely acts similarly to a telephone switchboard, conducting all thoughts which pass through it.

The last of the seven planes is the all-encompassing *Spiritual World.* Man's spiritual body is composed of the most high-frequency vibrations of all. By functioning in the spiritual body, one can get beyond time and space because spirit embraces all and flows uninterruptedly through everything. It is possible to contact the world of spirit within our microcosmic bodies because it is the life of the spiritual world which interpenetrates and sustains us.

Most texts on chakras discuss seven spiritual centers (chakras) in the human body that are visualized by people with extrasensory perception (i.e., psychics). Each chakra vibrates at a different speed and emits different colors to the clairvoyant eye. From observing the location of chakras, it has been confirmed that the seven areas are associated with specific organs in the body (Greenwell, 1990).

As one can see in Figure 12, the connections are associated with very important internal organs. Therefore, a malfunctioning organ or physiological system can cause diminished energy transformation. Since a more in depth examination of the entire chakra system would exceed the confines of this book, further reading in a more comprehensive book on chakras is recommended (Leadbeater, 1987; Bruyere, 1991). Our interest in mentioning the chakras is to show the reader that there are physiological systems in the body capable of channeling the energies of the universe. The chakra system is the junction site for the material-spiritual connection.

The chakras represent the multicommunication interface among the nervous system, the skin, and the spiritual world. Individuals who possess a degree of clairvoyance can see chakras radiate in the ethereal body where they are represented by saucer-like depressions or vortices on the surface. When quite undeveloped they appear as small circles about two inches in diameter, glowing dully in the ordinary man; but when awakened and vivified they are seen as coruscating whirlpools, much increased in size and resembling miniature suns (Leadbeater, 1987). Although we are saying that the nervous system and endocrine system function as interfaces, the chakras emanate through the ethereal body, not the physical body.

Two main points can be made about a relationship between melanin and chakras. First, the locations on the body in which chakras function are associated with the embryological origin of melanin. Recall that melanin is found in the neural crest of the developing human cell. Some of the derivatives that migrate from neural crest cells are the sensory nerves and ganglia which receive impulses from sense organs, autonomic ganglia, and the adrenal medulla. All of these body sites are critical for chakra functioning.

Secondly, melanin has been proposed to be effective as a device for radiation-less conversion of the energy of harmfully excited molecules into innocuous vibrational energy (McGinness and Proctor, 1973). The proposal by McGinness and Proctor is congruent with the way chakras channel cosmic energy in and out of the physical body. In addition, if melanin can be thought of as an amorphous semi-conductor (McGinness, 1972), then one can see how the conductive properties of melanin can increase the kinetic energy of electrons, propelling them into excited states. Melanin in the surface of the skin, in the internal organs and the nervous system, can enhance a person's connection to the spiritual world and act as a battery charger for the chakra system.

In conclusion, the chakra system is the human body's direct link to the spiritual realm. From the base of the spine to the crown of the head, a spiraling energy force rises to awake a person's spiritual consciousness. Although people are generally aware of their spirituality, proper meditation is the key to maximizing spiritual experience. Meditation can allow one to quietly dwell in the dark matter of the universe and be at one with the creator of this world. The dark matter of the universe and the blackness of melanin are the material planes that connect our mind and body to the spiritual world.

11

Prospective Research

In the previous chapters, I covered numerous topics related to melanin functioning. This final chapter will explore some behavioral disorders and physical abnormalities associated with melanin functioning. The information gathered from these disorders or abnormalities will help formulate research questions for African-centered scientists in their quest to optimize the health of people of African descent. Planning research projects regarding the black human and melanin as the central focal point may generate development of medical facts and social practices that will ensure good medical care for black people in the future (Barnes, 1988).

Advances in medicine have contributed to a longer life expectancy for humans. As people live longer, the elderly become increasingly vulnerable to neurodegenerative disorders. One neurodegenerative disorder that affects mostly the elderly is Parkinson's disease (PD). PD is pathologically characterized by destruction of dopaminergic cells in the midbrain region called the substantia nigra. The substantia nigra is a highly melinated subcortical structure. As part of the basal ganglia (motor system), it has an integral role in producing slow coordinated and deliberate movements. The special bioelectronic properties associated with melanin (see Chapter three) help to facilitate the conduction of nerve impulses. As a result, melanin in the brain can directly affect coordinated movement.

Besides destruction of substantia nigra neurons, there is also degeneration of cells in the ventral tegmental area and the locus coeruleus (Kaplan and Sadock, 1987). The combined effect of this neurodegeneration leads to tremor, muscle rigidity, bradykinesia (slow movements), stooped posture and a shuffling gait. By observing the psychomotor impairments of neurodegenerative disorders, one can infer how the brain controls normal behavior.

A thorough investigation of the behavioral impairments and motor deficits associated with PD can indirectly reveal how the substantia nigra functions. It can be speculated that a highly melinated substantia nigra can produce advanced motor skills throughout the lifetime of the organism. Future research should ascertain whether or not there are any differences in melanin content and dopamine production in the substantia nigra of specific ethnic groups. An answer to this question could reveal potential differences in brain structures containing neuromelanin in black and white patients. Interestingly, younger individuals are mysteriously developing PD. Future studies should be done to find the etiological factors that cause PD in young people.

Several advanced techniques could be utilized in studying brain regions in a living and functioning human. For example, computerized axial tomography or nuclear magnetic resonance imaging could be employed for scanning the brain for differences in the size of the substantia nigra. In addition, positron emission tomography could be used to study how much dopamine is being utilized by the substantia nigra during mental tasks requiring the individual to think about movement or to perform a simple motor task while in a stationary position.

There are other disorders associated with melanin besides brain deteriorating diseases. For example, albinism is a genetic defect that results in a direct loss of melanin production. Another genetic defect that indirectly decreases melanin production is phenylketonuria (PKU). PKU is transmitted as a simple recessive autosomal Mendelian trait and occurs in people of a predominantly North European origin (Kaplan and Sadock, 1987). In the United States, one to two percent of white people and almost no black people are heterozygous, or "carriers," for the condition. Although post-natal tests for PKU have greatly reduced the number of PKU cases, it is important that we understand the nature of the disorder in relation to melanin.

PKU is based on an abnormality in the body's metabolism with respect to phenylalanine, a common amino acid obtained from the diet. Normally, some phenylalanine is incorporated into proteins the body synthesizes, and most of the remainder is converted in the liver into tyrosine. Tyrosine, another essential amino acid, is critical for the conversion of both neurotransmitters and melanin. Children with PKU are born deficient in the enzyme, phenylalanine hydroxylase, that converts phenylalanine to tyrosine. As a result of the enzyme deficiency, children accumulate excessive levels of phenylalanine. Since phenylalanine and tyrosine are precursors of some of the brain's most important neurochemicals, the deficiency can cause devastating mental problems.

The majority of patients with PKU are severely retarded, but some are reported to have borderline or normal intelligence. Clinically, PKU children are hyperactive and erratic. Temper tantrums are common, and some patients display bizarre movements with their bodies and upper extremities, like the twisting of hands. Their behavior sometimes resembles that of autistic or schizophrenic children. PKU patients have poor coordination and perceptual difficulties, and their verbal as well as nonverbal communication is severely impaired or nonexistent (Kaplan and Sadock, 1987).

In addition to mental problems, PKU patients have changes in pigmentation. The color change is due to a lack of tyrosine needed for the biosynthesis of melanin. Consequently, PKU children are usually light colored with blond hair and blue eyes (Kalat, 1984). In short, a single enzyme deficiency leads directly to a wide variety of behavioral as well as physical abnormalities.

Evidence that deficient levels of tyrosine can cause both skin changes and behavioral disturbances suggests that skin melanin and neuromelanin are formed by the same biosynthetic pathway. Although there is experimental evidence to indicate that skin melanin and neuromelanin are not formed by similar pathways (Rogers and Curzon, 1975), the physical evidence found in PKU patients seems to indicate that there is a similar pathway. It appears to be a complex issue that deserves further investigation. To provide an explanation at this point without debating over the biosynthesis of skin melanin and neuromelanin, it may be more appropriate to state that neuromelanin is genetically programmed to function at a different capacity depending upon the individual's overall genetic capacity to produce melanin.

There are three specific research questions that could be asked in relation to melanin and PKU: First, is the direct loss of neuromelanin the primary factor in the cause of mental retardation, loss of coordination, and other behavioral disturbances in PKU patients? Second, are the low levels of neurotransmitters the primary cause of behavioral disturbances? Third, are the behavioral disturbances simply a result of high levels of phenylalanine? These questions could help validate the key role of melanin during fetal and early childhood development. Since PKU patients can be treated with dietary supplements, the testing for the proposed questions would need to be done during the critical stage before treatment and recovery. Ostensibly, studies of this type would create some ethical concerns.

There are additional disorders in humans resulting in unusual pigmentation which should also be researched. Some are associated with complex endocrine feedback loops such as Addison's disease and Cushing's syndrome. In cases of Addison's disease, patients have decreased steroid hormone production from the adrenal cortex, resulting in an increased output of ACTH and MSH from the pituitary gland to stimulate a rise in melanin synthesis in the skin. With respect to Cushing's syndrome, excessive amounts of ACTH and POMC-derived melanotropic peptides are secreted to cause hyperpigmentation. The physical changes associated with Addison's and Cushing's diseases could shed light on the central brain mechanisms involved in stimulating skin pigmentation. We are well aware of the sun's influence on skin pigmentation; however, brain mechanisms stimulating skin pigmentation are not so well understood. Once again, abnormal processes can help us determine how the brain controls normal bodily functions.

Two depigmenting disorders that can be genetically transmitted are piebaldism and vitiligo. Piebaldism, or "partial albinism," is an inherited absence of melanocytes due to faulty differentiation or improper migration of melanin cells in the neural crest. Vitiligo is the most prevalent pigmentary disorder occurring throughout the animal kingdom and one-to-two percent of humans. Vitiligo is characterized by a progressive loss of melanocytes at some time during the life-span of the organism.

In terms of prospective research, investigations into these pigmentary disorders should lead to eventual cures for skin cancer. If one can study the nature of the defects in melanoma cells, one can develop potential treatments to block the progression of deadly skin cancer. Currently, the *Pigment Cell Research* journal reports findings from such research. In fact, scientists have already developed pharmacological agents that can stimulate melanin production or cause depigmentation. These substances can be found in any *Physician's Desk Reference.*

Melanin has an established role as a natural sunscreen, however, there are additional functions that can be uncovered by studying the deficits associated with the aforementioned biological disturbances. It is not simple to inject melanin into humans and study the resultant effects, so investigating the deficits can indirectly provide information on the role of melanin in the skin and nervous systems. An analysis of deficits associated with melanin dysfunctioning can be useful but cannot entirely explain the numerous roles of melanin.

In terms of external melanin, African-centered scientists should investigate diets recommended for enhancing the appearance of skin. Proper nutrition can make the skin smooth and lustrous. Diets rich in vitamin A, beta carotene, and water clean out impurities in the system and add vitality to the skin. As a matter of fact, Western scientists are

presently marketing ointments that contain vitamin A and beta carotene.

Rather than depend on externally applied substances for improving health, our diets should consist of nothing but natural substances
to supply our needs. In this modern age of processed and packaged
food items, it is difficult to structure a diet of solely natural foods. It
takes little time, however, to read ingredients to determine if a food
substance is natural or if it contains artificial substances. If the
cumulative effects of artificial food items can lead to neurodegenerative
disorders such as PD and Alzheimer's disease, then diminishing the
potential development of neurodegenerative diseases ought to be
reason enough for one to be aware of the ingredients in one's food.

Unfortunately, in our materialistic culture, most people probably
take better care of their car than their own body. For example, gas, oil,
and antifreeze are the main fluids needed for the proper functioning of
a car. No artificial chemicals imitating gas, oil, or antifreeze should be
used because the car will malfunction. A car is a vehicle for
transportation, but one's body is a vehicle for transformation. And
while it is important to know what kind of gas, oil, etcetera is going
into one's car, it is all the more important to be aware of what we put
into our bodies.

Diet can have a significant impact on behavior. The
neurodegenerative disorders mentioned earlier, and the genetic abnormalities associated with PKU, can be positively as well as negatively influenced by diet. As a matter of fact, all the vitamins and
minerals obtained from the diet are extremely necessary to the body's
internal chemistry. A description of specific locations in which
vitamins affect the biosynthesis of neurotransmitters can be found in
Figure 13.

Figure 13 demonstrates the role of vitamins C and B6 in the
biosynthesis of specific neurotransmitters. Melanin formation is
dependent upon these vitamins as well as amino acids, such as
tyrosine. African-centered nutritionists should research the right
combinations of food to enhance melanin functioning. Andrews
(1990) has provided some recipes in his book, and researchers from
Khem-Sci Nutrition, Inc. have developed a powdered formula called

PROMELANIN 2000. PROMELANIN 2000 is one of many products scientifically designed to meet the nutritional biochemistry requirements for people of African descent.

In addition to nutrition, further scientific research should be performed in the area of developmental embryology. For instance, how is melanin assisting cells to migrate to their destination sites? Is melanin more pronounced during different stages of fetal development? Also, what is the significance of the dark melinated line formed on the center of a pregnant mother's stomach, extending from the middle of the chest to the vaginal area? Is there an association between melanin formation during pregnancy and enlarged anterior pituitary glands? These and many more developmental embryology questions could be asked.

After the African child comes out of the womb, how can African parents mold, strengthen, and capitalize on the advanced sensorimotor development (Wilson, 1978; 1991)? African children may need a different style of learning (Bell, 1994) to compensate for their advanced development. Developmental psychologists should implement appropriate theoretical models to maximize school performance of African children rather than have them attend schools that would label them hyperactive and learning impaired.

In addition to developmental psychologists, neuro-psychologists and cognitive psychologists can find many research ideas in the study of melanin and melanin's effect on human potential. For example, it is known that sufficient amounts of melanin in the retina can provide increased visual acuity. Neuropsychologists and cognitive psychologists, therefore, could study visual processing and learning styles. There might be specific learning styles that could be developed to increase one's effectiveness in the learning process —speed reading, for example.

In addition to the visual system, researchers can investigate the specific effects of melanin in the auditory system. Some nonwestern cultures honor ancestors and express the hearing of voices. The hearing of voices is not linked to a specific psychological disorder in these nonwestern cultures. Although Western science labels auditory hallucinations as a symptom of psychotic behavior, the hearing of voices associated with other cultures may relate to melanin function-ing and the expanding of consciousness. Honoring ancestors and communicating with the spiritual world are phenomena associated with nonwhite people. If it is natural and not "abnormal" for other cultures to experience auditory hallucinations, then it may be due to extrasensory perception.

In terms of "abnormal" behavior, high amounts of dopamine are found in patients suffering from psychotic symptoms. Therefore, hallucinations are primarily the result of brain dysfunctioning rather than disturbances in the inner ear. If the brain mechanisms linked to hallucinations are a result of high levels of brain dopamine, then there may be some involvement with neuromelanin. If this dopamine is functioning in melinated brain regions such as the substantia nigra, it may influence the motor responses that may occur during specific hallucinatory experiences.   In comparison to psychotic symptoms, researchers may also attempt to investigate individuals who meditate and experience altered states of consciousness. Brain scans could be used to study the cerebral blood flow of the brain during metaphysical experiences. The cerebral blood flow associated with metaphysical experiences could be correlated with the cerebral blood flow of psychotic patients when they are experiencing auditory hallucina-tions. During either experience, one may ask if there is increased blood flow in highly melinated brain regions.

Furthermore, African-centered melanin researchers could inves-tigate inner ear melanin and the effects of music. For example, music definitely has an influence on behavior and does so by vibrations. How does the "thumping" bass of black music affect highly melinated individuals versus individuals who lack melanin? Can differences in inner ear melanin have a significant impact on the type of music people create and listen to? The drum is an ancient instrument in African culture. Why has it been so important to African people in terms of communication? Since some say that the drum can actually heal, how can we devise experiments to validate the healing effects scientifically?

Historically, one should take note that white slave masters barred the drum in America during the enslavement process. Did the drum possess significant properties that motivated highly melanized Africans to fight against oppression during enslavement? Research should be conducted to investigate the vibratory effects of drumming on human behavior and specific body sites such as the inner ear.

Today, the drum is used in all forms of Black music. Rap music, in particular, is an African phenomena that deserves research. Rapping to music is no easy task and requires advanced brain functioning. Moving rhythmically, staying on beat, thinking, creating and remembering words is a skill mastered by very few outside the black community. The subcortical brain structures in the basal ganglia that are responsible for movement and the intercommunication between the two hemispheres, must function at a higher level of brain functioning in order to produce a musical form such as rap. Studies using the electroencephalograph could be conducted to measure the brain activity of skilled versus nonskilled rappers.

On a grand scale, there is no simple way to study all melanin. There are high tech space shuttles and telescopes, but African-centered scientists, in contrast to Western (white) scientists, are less concerned with billion dollar explorations in outerspace. Here on earth, feasible experiments can be done to investigate melanin's ability to optimize the personal search for divine universal consciousness. The bioelectronic properties of melanin in our bodies can be studied to ascertain how energy can be transformed from dark matter into bodily material. By meditating and linking up with the energy in the universe, one can obtain unlimited mind power.

African-centered melanin researchers should invest in melanin research as we head for the next century. This exploration and investigation will require more than armchair commitment, however. High tech laboratory equipment will be needed if we are to uncover the science of melanin. Hopefully, this book has stimulated enough interest to bring about new research, that which will focus on ways to optimize our lives in today's world.

The environment is under assault from continuing ecological disasters. And the quality of food has declined. As people of different cultures move closer in proximity, new strains of diseases and illnesses are being developed. To maintain a healthy lifestyle and avoid the adverse effects of a polluted environment, low quality food, and strange diseases, melanin functioning can help to optimize health and slow the aging process. Rather than focus on the quantity of life in terms of how long one lives, melanin research should be conducted to improve the quality of health and life.

References

Abe, K., Island, D.P., Liddle, G.W., Fleisher, N. and Nicholoson. R. (1967). Radioimmunologic evidence for alpha-MSH (melanocyte stimulating hormone) in human pituitary and tumor tissue. *Journal of Clinical Endocrinology and Metabolism*, 27, 46-52.

Alder, V.S. (1968). *The Finding of the Third Eye.* Maine: Samuel Weiser, Inc. Originally published 1938.

Amaral, D.G. and Sinnamon, H.M. (1977). The locus coeruleus: neurobiology of a central noradrenergic nucleus. *Progress in Neurobiology*, 9, 147-196.

Amen, N.A. (1993). *The Ankh: African Origin of Electromagnetism.* Jamaica, NY: Nur Ankh Amen Press.

References

Andrews, M. (1989). *Color Me Right...Then Frame Me in Motion.* Tennessee: Seymour-Smith Inc.

Barden, H. (1969). The histochemical relationship of neuromelanin and lipofuscin. *Journal of Neuropathology and Experimental Neurology*, 28, 419-441.

Barnes, C. (1988). *Melanin: The Chemical Key to Black Greatness, Vol.l.* Houston, Texas: C.B. Publishers.

Barnes, C. (1993). *Jazzy Melanin: A Novel.* Houston, Texas: Melanin Technologies.

Barr, F.E. (1983). Melanin: the organizing molecule. *Medical Hypothesis*, 11(1).

Bayley, N. (1965). Comparisons of mental and motor test scores for ages 1-15 months by sex, birth order, race, geographic location and education of parents. *Child Development*, 36, 379-410.

Bazelon, M., Fenichel, G.M. and Randall, J. (1967). Studies on neuromelanin. I. a melanin system in the human adult brainstem. *Neurology*, 17, 512-519.

Bell, Y.R. (1994). A culturally sensitive analysis of black learning style. *The Journal of Black Psychology*, 20(1), 47-61.

Ben-Jonathan, N. and Peters, L.L. (1982). Posterior pituitary lobectomy: differential elevation of plasma prolactin and luteinizing hormone in estrous and lactating rats. *Endocrinology*, 110, 1861-1865.

Ben-Jonathan, N. (1985). Dopamine: a prolactin-inhibiting hormone. *Endocrinology Review*, 6, 564-589.

Blois, M.S. (1969). In F. Urbach (ed.), *The Biological Effects of Ultraviolet Radiation*. New York: Pergamon.

Blois, M.S. (1971). In T. Kawamura, T.B. Fitzparick, M. Seiji (Eds.), *Biology of Normal and Abnormal Melanocytes*. Baltimore: University Press.

Bogerts, B. (1981). A brainstem atlas of catecholaminergic neurons in man, using melanin as a natural marker. *Journal of Comparative Neurology*, 197, 63-80.

Bonaccorsi, P. (1965). Li colore dell'iride come 'Test' di valutazione quantitative, nell'uomo, della concentrazzione di melania nella stria vascolare. *Annals in Laryngology*, 64, 725-738.

Bourquelot, E. and Bertrand, A. (1895). Le bleuissement et le noircissement des champignons. *Current Reviews in Soc. Biology*, 47, 582-584.

Bover, A., Hadley, M.E. and Hruby, V.J. (1974). Biogenic amines and control of melanophore stimulating hormone release. *Science*, 184, 70-72.

Bradley, M. (1991). *The Iceman Inheritance: Prehistoric Sources of Western Man's Racism, Sexism and Aggression*. New York: Kayode Publications Ltd. (originally published 1978).

Breathnach, A.S. (1988). Extra-cutaneous melanin. *Pigment Cell Research*, 1, 234-237.

Bruyere, R.L. (1991). *Wheels of Light: A Study of the Chakras* (2nd ed.). California: Bon Productions.

Calne, D.B. (1991). Neurotoxins and degeneration in the central nervous system. *Neurotoxicology*, 12, 335-340.

References

Celio, M.R. (1979). Distribution of beta-endorphin immunoreactive cells in human fetal and adult pituitaries and pituitary adenomas. *Journal of Histochemical Cytochemistry*, 27, 1215.

Clark (X), C., McGee, D.P. Nobles, W. and Weems, L. (1975). Voodoo or I.Q.: an introduction to African psychology. *Journal of Black Psychology*, 1(2), 9-19.

Clark, D., Thody, A.J., Shuster, S. and Bowers, H. (1978). Immunoreactive alpha-MSH in human plasma in pregnancy. *Nature*, 273, 163-164.

Commoner, B., Townsend, J. and Pake G.E. (1954). Free radicals in biological materials. *Nature*, 174, 689-691.

Corti, A. (1851). Recherches surl'organe de l'ouic mammiferes. *Premiere partie. 2 wiss Zoologie*, 3, 109-169.

Creel, D., O'Donnell, F.E. and Witkop, C.J. (1978). Visual system anomalies in human ocular albinos. *Science*, 201, 931-933.

Creel, D., Garber, S.R., King, R.A. and Witkop, C.J. (1980). Auditory brainstem anomalies in human albinos. *Science*, 209, 1253-1255.

Culp, C.H., Eckels, D.E. and Sidles, P.H. (1975). *Applied Physics,* 46, 3658-3659.

Curzon, G. (1975). Metals and melanins in the extrapyramidal centers. *Pharmacological Therapeutics Bulletin*, 1(4), 673-684.

Davis, M.D. (1986). The hypothalamo-hypophyseal rat explant in vitro: endocrinological studies of the pars intermedia dopaminergic neuronal input. *Journal of Physiology*, 370, 381-393.

Diop, C.A. (1973). Pigmentation of the ancient egyptians: tests by melanin analysis. *Bulletin de L'Institut fondamental D'Afrique Noire*, Serie B, Science Humaines, Tome XXXV, No.3, Julliet, 551-530.

Diop, C.A. (1974). *The African Origin of Civilization: Myth or Reality.* Lawrence Hill and Co., Publishers Inc.

Diop, C.A. (1991). *Civilization or Barbarism: An Authentic Anthropology.* New York: Lawrence Hill Books

Drake, S.C. (1987). *Black Folk Here and There, Vol. 1*, UCLA: Center of Afro-American Studies.

Elmer-Dewitt, P. (1989, January). Preparing for the worst. *Time*, pp. 70-71.

Epstein, J.H. (1990). Phototherapy and photochemotherapy. *The New England Journal of Medicine*, 322(16), 1149-1151.

Feldman, R.S. and Quenzer, L.F. (1984). *Fundamentals of Neuropsychopharmacology.* Sunderland, MA: Sinauer Associates, Inc.

Fenichel, G.M. and Bazelon, M. (1968). Studies on neuromelanin. II. melanin in the brainstems of infants and children. *Neurology*, 18, 817-820.

Filatovs, J., McGinness, J. and Corry, P. (1976). Thermal and electronic contributions to switching in melanins. *Biopolymer*, 15, 2309-2312.

Finch, C. (1990). The evolution of the caucasoid. In I. Van Sertima (ed.), *African Presence in Early Europe: Journal of African Civilization,* 7(2), (pp. 17-22). New Brunswick, NJ: Transaction Publishers.

Finch, C. (1991). *Echoes of the Old Darkland*. Decatur, GA: Khenti, Inc.

Fitzpatrick, T.B., Hopkins, C.E., Blickenstaff, D.D. and Swift, S. (1955). Augmented pigmentation and other responses of normal human skin to solar radiation following oral administration of 8-methoxypsoralen. *Journal of Investigative Dermatology*, 25, 187-190.

Foley, J.M. and Baxter,D. (1958). On the nature of pigment granules in the cells of the locus coeruleus and substantia nigra. *Journal of Neuropathology and Experimental Neurology*, 17, 586-598.

Forbes, J. (1992). *Columbus and Other Cannibals*. New York: Autonomedia.

Fuller, N. (1969). *Textbook for Victims of White Supremacy*. Library of Congress.

Gan, E.V., Haberman, H.F. and Menon, I.A. (1976). Electron transfer properties of melanin. *Archives in Biochemistry and Biophysics,* 173, 666-672.

Gan, E.V., Lam, K.M., Haberman, H.F. and Menon, I.A. (1977). Electron transfer properties of melanins. *British Journal of Dermatology*, 96, 25-28.

Geber, M. (1958). The psychomotor development of african children in the first year and the influence of maternal behavior. *Journal of Social Psychology*, 47, 185-195.

Golob, R. and Brus, E. (Eds.) (1990). *The Almanac of Science and Technology: What's New and What's Known*. Orlando, Florida: Harcourt Brace Jovanovich, Publishers.

Gould, S.J. (1981). *The Mismeasure of Man*. New York: W.W. Norton and Company.

Graham, D.G. (1979). On the origin and significance of neuromelanin. *Archives of Pathology and Laboratory Medicine*, 103, 359-362.

Granger, J. (1988). *Mo'*. Washington, D.C.: Uraeus Publishing, Inc.

Greenwell, B. (1990). *Energies of Transformation: A Guide to the Kundalini Process*. Shakti River Press.

Griuale, M. and Dieterlen, G. (1986). *The Pale Fox*. Arizona: Continuum Foundation. Originally published 1965.

Grota, L.J., Lewy, A.J., Goldsmith, L.A. and Brown, G.M. (1985). Psoralen increases melatonin levels without ultraviolet light. The medical and biological effects of light. In R.J. Wurtman, M.J. Baum and J.T. Potts (Eds.), *Annals of The New York Academy of Sciences*, Volume 453, (pp. 385-387).

Gudelsky, G.A. (1981). Tuberoinfundibular dopamine neurons and the regulation of prolactin secretion. *Psychoneuroendocrinolgy*, 6, 3-16.

Guillery, R.W., Hickey, T.L., Kass, J.H., Felleman, D.L., Debruyn, E.J. and Sparks, D.L. (1984). Abnormal central visual pathways in the brain of an albino green monkey (Cercopithecus aethiops). *Journal of Comparative Neurology*, 226, 165-183.

Guillery, R.W. (1986). Neural abnormalities of albinos. *Trends in Neuroscience*, 9, 364-367.

Guyton, A.C. (1982). *Human Physiology and Mechanisms of Disease* (3rd ed.). Philadelphia: W.B. Saunders Company.

Hadley, M.E. (1972). The significance of vertebrate integumental pigmentation. *American Zoology*, 12, 63-76.

References

Hadley, M.E. and Hruby, V.J. (1977). Neurohypophysial peptides and the regulation of melanophore stimulating hormone (MSH) secretion. *American Zoology*, 17, 809-821.

Hadley, M.E. (1988). *Endocrinology*(2nd ed.). New Jersey: Prentice Hall.

Harber, L.C. and Bickers, D. (1989). *Photosensitivity Diseases: Principles of Diagnosis*. Toronto: B.C. Decker.

Herlant, M. and Pasteels, J.L. (1967). Histophysiology of the human anterior pituitary. *Methods Achieved in Experimental Pathology*, 3, 250-305.

Hilding, D.A. and Ginsberg, R.D. (1977). Pigmentation of the stria vascularis. *Acta Ootolaryngology*, 84, 24-37.

Holick, M.F. (1985). The photobiology of vitamin D and its consequences for humans. In R.J. Wurtman, M.J. Baum and J.T. Potts (Eds.). The medical and biological effects of light, *Annals of The New York Academy of Sciences, Volume 453*, (pp. 1-13).

Jackson, F. and Jackson, J. (1978). *Infant Culture*. New York: New American Library.

Jackson, S. and Lowry, P.J. (1983). Secretion of pro-opiocortin peptides from isolated perfused rat pars intermedia cells. *Neuroendocrinology*, 37, 248-257.

Jaroff, L. (1994, April 4). Teaching reverse racism. *Time*, pp. 74-75.

Kalat, J.W. (1984). *Biological Psychology* (2nd ed.). Belmont, California: Wadsworth Publishing Company.

Kaplan, H.I. and Sadock, B.J. (1988). *Synopsis of Psychiatry: Behavioral Sciences Clinical Psychiatry* (5th ed.). Baltimore: Williams and Wilkins.

Kastin, A.J., Gennser, G., Arimura, A., Miller, M.C. and Schally, V. (1968). Melanocyte-stimulating and corticotrophin activities in human foetal pituitary glands. *Acta Endocrinologia*, 58, 6-10.

King, R. (1990). *African Origin of Biological Psychiatry.* Tennessee: Seymour-Smith Inc.

King, R. (1994). *Melanin: The Key to Freedom.* Hampton, VA: United Brothers and United Sisters Communications Systems, Inc.

Lacy, M.E. (1984). Phonon-electron coupling as a possible transducing mechanism in bioelectronic processes involving neuromelanin. *Journal of Theoretical Biology*, 111, 201-204.

LaFerriere, K.A., Arenberg, I.K., Hawkins, J.E. and Johnson, L.G. (1974). Melanocytes of the vestibular labyrinth and their relationship to the microvasculature. *Annals in Otology*, 83, 685-694.

Larsson, B. and Tjalve, H. (1979). Studies on the mechanisms of drug-binding to melanin. *Biochemical Pharmacology*, 28, 1181-1187.

Leadbeater, C.W. (1987). *The Chakras* (5th ed.). Illinois: The Theosophical Publishing House. Originally published 1927.

Leiderman, G. (1973). African infant precocity and some influences during the first year. *Nature*, 242, 247-249.

Lemonick, M.D. (1989, January). Feeling the heat. *Time*, pp. 42.

References

Lerner, A.B., Case, J.D., Takahashi, Y, Lee, T.H. and Mori, W. (1958). Isolation of melatonin, the pineal gland factor that lightens melanocytes. *Journal of American Chemical Society*, 80, 2587.

Lerner, A.B. and McGuire, J.S. (1961). Effect of alpha- and beta-melanocyte-stimulating hormones on the skin colour of man. *Nature*, 189, 176-179.

Lerner, A.B. and McGuire, J.S. (1964). Melanocyte-stimulating hormone and adrenocorticotrophic hormone. *New England Journal of Medicine*, 270, 539-546.

Lichtensteiger, W. and Lienhart, R. (1977). Response of mesencephalic and hypothalamic dopamine neurons to alpha-melanocyte stimulating hormone: mediated by area postrema? *Nature*, 266, 635-637.

Lichtensteiger, W. and Monnet, F. (1979). Differential response of dopamine neurons to alpha-melanotropin and analogues in relation to their endocrine and behavioral potency. *Life Science*, 25, 2079-2087.

Lichtensteiger, W. and Schlumpf, M. (1986). Permanent alteration of peptide feedback on dopamine neurons after injection of alpha-melanotropin antiserum at a critical period of postnatal development. *Brain Research*, 368, 205-210.

Lillie, R.D. (1955). The basophilia of melanins. *Journal of Histochemical Cytochemistry*, 3, 453.

Lillie, R.D. (1957). Metal reduction reactions of the melanins: histochemical studies. *Journal of Histochemical Cytochemistry*, 5, 325-333.

Lindley, S.E., Lookingland, K.J. and Moore, K.E. (1990). Activation of tuberoinfundibular but not tuberohypophysial dopaminergic neurons following intracerebroventricular administration of alpha-melanocyte-stimulating hormone. *Neuroendocrinology*, 51, 394-399.

Lindquist, N.G., Larsson, B.S. and Lyden-Sokolowski, A. (1987). Neuromelanin and its possible protective and destructive properties. *Pigment Cell Research*, 1, 133-136.

Lyttkens, L., Larsson, B., Stahle, J. and Engleson, S. (1979). Accumulation of substances with melanin affinity to the internal ear. *Advances in Otorhinolaryngology*, 25, 17-25.

Mann, D.M.A. and Yates, P.O. (1983). Possible role of neuromelanin in the pathogenesis of parkinson's disease. *Mechanisms in Age Development*, 21, 193-203.

Marsden, C.D. (1983). Neuromelanin and parkinson's disease. *Journal of Neural Transmission*, 19, 121-141.

Marwan, M.M., Jang J., Castrucci, A.M. and Hadley, M. (1990). Psoralen stimulates mouse melanocyte and melanoma tyrosinase activity in the absence of ultraviolet light. *Pigment Cell Research*, 3, 214-221.

McGinness, J. (1972). Mobility gaps: A mechanism for band gaps in melanin. *Science*, 177, 196-197.

McGinness, J. and Proctor, P. (1973). The importance of the fact that melanin is black *Journal of Theoretical Biology*, 39, 677-688.

McGinness, J., Corry, P. and Proctor, P. (1974). Amorphous semiconductor switching in melanins. *Science*, 183, 853-855.

McGinness, J. (1985). A new view of pigmented neurons. *Journal of Theoretical Biology*, 115, 475-476.

References

Meyer zum Gottesberge, A.M. (1988). Physiology and pathophysiology of inner ear melanin. *Pigment Cell Research*, 1, 238-249.

Moses, H.L., Ganote, C.E., Beaver, D.L. and Schuffman, S.S. (1966). Light and electron microscopic studies of pigment in human and rhesus monkey substantia nigra and locus coeruleus. *Anatomical Record*, 155, 167-184.

Muramoto, N.B. (1988). *Natural Immunity: Insights on Diet and Aids*. California: George Ohsawa Macrobiotic Foundation.

Nobles, W. (1986). *African Psychology*. California: Black Family Institute.

Nsamba, C. (1972). A comparative study of the etiology of vertigo in the african. *Journal of Laryngology and Otology*, 86, 917-925.

O'Donohue, T.L. and Dorsa, D.M. (1982). The opiomelanotropinergic neuronal and endocrine systems. *Peptides*, 3, 353-395.

Olszewski, J. (1964). *Cytoarchitecture of the Human Brain*. New York: Stern and Birjelow.

Oppenheimer, S,B. and Lefevre, G. (1984). *Introduction to Embryonic Development*. Massachusetts: Allyn and Bacon, Inc.

Ortiz de Montellano, B. (1993). Melanin, afrocentricity, and pseudoscience. *Yearbook of Physical Anthropology*, 36, 33-58.

Pawelek, J.M and Korner, A.M. (1982). The biosynthesis of melanin. *American Scientist*, 70, 136-145.

Penny, R.J. and Thody, A.J. (1978). An improved radioimmunoassay for alpha-melanocyte stimulating hormone in the rat: serum and pituitary alpha-melanocyte stimulating hormone levels after drugs which modify catecholamine neurotransmission. *Neuroendocrinology*, 25, 193-203.

Pierce, J. (1980). *Magical Child: Rediscovering Nature's Plan for Our Children*. New York: Bantam Books.

Poole, R. (1989a). Ecologists flirt with chaos. *Science*, 243, 310-313.

Poole, R. (1989b). Quantum chaos: enigma wrapped in a mystery. *Science*, 243, 893-895.

Poole, R. (1989c). Chaos theory: how big an advance? *Science*, 245, 26-28.

Proctor, P., McGinness, J. and Corry, P. (1974). A hypothesis on the preferential destruction of melanized tissues. *Journal of Theoretical Biology*, 48, 19-22.

Purves, H.D. (1966). Cytology of the hypophysis. In G.W. Harris and B.T. Donovan (Eds.). *The Pituitary Gland, Vol. 1*, (pp.147-232). London: Butterworths.

Quevedo, W.C., Fitzpatrick, T.B., Pathak, M.A. and Jimbow, K. (1975). Role of light in human skin color variation. *American Journal of Physical Anthropology*, 43, 393-408.

Quevedo, W.C., Fitzpatrick, T.B. and Jimbow, K. (1985). Human skin color: origin, variation and significance. *Journal of Human Evolution*, 14, 43-56.

Ranvier, L. (1875). Labyrinthe membraneux et terminaison du nerv auditiv. In *Traite technique d'histologie* (pp.986-1020). Paris: Savy.

References

Redgrove, P. (1987). *The Black Goddess and the Unseen Real: Our Unconscious Senses and Their Uncommon Sense.* New York: Grove Press.

Rendell, P. (1979). *Introduction to the Chakras.* London: The Aquarian Press.

Rodgers, A.D. and Curzon, G. (1975). Melanin formation by human brain in vitro. *Journal of Neurochemistry,* 24 1123-1129.

Ronan, C. (1973). *Lost Discoveries.* MacDonald Publishing.

Sagan, C. (1980). *Cosmos.* New York: Random House.

Schell, J. (1982). *The Fate of the Earth.* New York: Avon Books.

Schizume, K., Lerner, A.B. and Fitzpatrick, T.B. (1954). In vitro bioassay for the melanocyte stimulating hormone. *Endocrinology,* 54, 553-560.

Schneider, W.C., Shelton, E. and Kuff, E.L. (1975). Association of DNA with melanin granules. *Journal of the National Cancer Institute,* 55(3), 665-669.

Schwaller de Lubicz, I. (1981). *The Opening of the Way: A Practical Guide to the Wisdom Teachings of Ancient Egypt.* Rochester, VT: Inner Traditions International.

Schwaller de Lubicz, R. A. (1977). *The Temple in Man.* Rochester, VT: Inner Traditions International.

Silman, R.E., Chard, T., Landon, J., Lowry, P.J., Smith, I. and Young I.M. (1977). ACTH and MSH peptides in the human adult and fetal pituitary gland. *Frontiers in Hormone Research,* 4, 179-187.

Smart, M. and Smart, R. (1972). *Children: Development and Relationships* (2nd ed.). New York: Macmillan.

Smotherman, W.P. and Robinson, S.R. (1990). The prenatal origins of behavioral organization. *Psychological Science,* 1(2), 97-105.

Smotherman, W.P. and Robinson, S.R. (1988). Dimensions of fetal investigation. In W.P. Smotherman and S.R. Robinson (Eds.), *Behavior of the Fetus* (pp. 19-34). Caldwell, NJ: Telford Press.

Souetre, E.J., Salvati, E., Belugou, J.L. et al. (1987). 5-methoxypsoralen increases the plasma melatonin levels in humans. *Journal of Investigative Dermatology,* 89, 152-155.

Souetre, E.J., Salvati, E., Belugou, J.L., Robert, P., Brunet, G. and Darcourt, G. (1988). Antidepressant effect of 5-methoxypsoralen: a preliminary report. *Psychopharmacology,* 95, 430-431.

Souetre, E.J., Salvati, E., Krebs, B., Belugou, J.L. and Darcourt, G. (1989). Abnormal melatonin response to 5-methoxypsoralen in dementia. *American Journal of Psychiatry,* 146, 1037-1040.

Souetre, E.J., De Galeani, B., Gastaud, P., Salvati, E. and Darcourt, G. (1989). 5-methoxypsoralen increases the sensitivity of the retina to light in humans. *European Journal of Clinical Pharmacology,* 36, 59-61.

Souetre, E.J., Salvati E., Belugou, J.L., Krebs and Darcourt, G. (1990). 5-methoxypsoralen as a specific stimulating agent of melatonin secretion in humans. *Journal of Clinical Endocrinology and Metabolism,* 71(3), 670-674.

Stewart, N. (1981). *Sensori-motor abilities of the African-American infant: implications for developmental screening.* Doctoral dissertation, George Peabody College for Teachers of Vanderbilt.

Stumpf, W.E. (1988). The endocrinology of sunlight and darkness: complimentary roles of vitamin D and pineal hormones. *Naturwissenschaften, 75,* 247-251.

Stumpf, W.E. and Privette, T.H. (1989). Light, vitamin D and psychiatry: role of 1,25 dihydroxyvitamin D3 (soltriol) in etiology and therapy of seasonal affective disorder and other mental processes. *Psychopharmacology, 97,* 285-294.

Swaab, D.F. and Visser, M. (1977). A function for alpha-MSH in fetal development and the presence of an alpha-MSH-like compound in nervous tissue. *Frontiers in Hormone Research, 4,* 170-178.

Swaab, D.F. and Martin, J.T. (1981). Functions of alpha-melanotropin and other opiomelanocortin peptides in labor, intrauterine growth and brain development. In D. Evered and G. Lawrenson (Eds.), *Peptides of the Pars Intermedia,* CIBA Foundation Symposium 81, (pp. 196-217). Bath England: Pitman Medical.

Tomita, Y., Fukushima, M. and Tagami, H. (1986). Stimulation of melanogenesis by cholecalciferol in cultured human melanocytes: a possible mechanism underlying pigmentation after ultraviolet irradiation. *Tohoku Journal of Experimental Medicine, 149,* 451-452.

Tota, G. and Bocci, G. (1967). Importance of the color of the iris in the evaluation of resistance of hearing to fatigue. *Reviews in Otoneuroophtalmology, 43,* 183-192.

Trefil, J. (1993). Dark matter. *Smithsonian, 24*(3), 27-35.

Van Sertima, I. (Ed.) (1987). *Great African Thinkers, Vol. 1.* New Brunswick, NJ: Transaction Books.

Van Woert, M.H. (1968). DPNH oxidation by melanin: inhibition by phenothiazines. *Proceedings in Social, Experimental and Biological Medicine*, 129, 165-171.

Velikovsky, I. (1977). *Earth in Upheaval*. New York: Pocket Books. Originally published 1955.

Velikovsky, I. (1977). *Worlds in Collision*. New York: Pocket Books. Originally published 1950.

Voltolini (1860). Anatomische und pathologisch-anatomische untersuchungen des gehororganes nebst funf sektions-fallen. *Archive Pathologie Anatomie und Physiologie. und fur klin. Medicin*, 27, 34-50.

Waldeyer, W. (1871). Hornerv und Schnecke. In *Stricker Handbuch der Lehre von den Geweben des Menschen und der Tiere* (pp. 913-963).

Walters, C. (1967). Comparative development of negro and white infants. *Journal of Genetic Psychology*, 110, 243-251.

Walther, T., Haustein, U.F., Rytter, M. and Gast W. (1989). Effect of 8-methoxypsoralen on granulocytes without ultraviolet irradiation. *Photodermatology*, 6, 185-187.

Warnette, K. and Andrews, M. (1990). Melanin consciousness: symbolic of the future. *The Malachi Papers*, January 15, Vol. 1, p. 2.

Welsing, F.C. (1991). *The Isis Papers: The Keys to the Colors*. Chicago: Third World Press.

West J.A. (1987). *Serpent in the Sky: The High Wisdom of Ancient Egypt*. New York: Julian Press.

References

Wilson, A.N. (1978). *The Developmental Psychology of the Black Child*. New York: Africana Research Publications.

Wilson, A.N. (1991). *Awakening the Natural Genius of Black Children*. New York: Afrikan World InfoSystems

Witkop, C.J., Quevedo, W.C. and Fitzpatrick, T.B. (1983). Albinism and other disorders of pigment metabolism. In J.B. Stansbury, J.B. Wyngaarden, D.S. Fredrickson, J.L. Goldstein and M.S. Brown (Eds.), *The Metabolic Basis of Inherited Disease* (5th ed.) (pp.301-346). New York: McGraw-Hill.

Wolf, D. (1931). Melanin in the inner ear. *Archives in Otolaryngology*, 14, 195-211.

Index

JAME RICHARD JUN